SPACESHIPS
OF THE
MIND

Overleaf. *'Canst thou bind the sweet influence of the Pleiades, or loose the bands of Orion?' God demanded of Job. The open star cluster of the Pleiades is seen by modern astronomers to consist of hundreds of stars, including the bright blue stars illustrated. It lies about 400 light-years away. Human interstellar expansion could put our descendants among the Pleiades in 25,000 years – the interval that separates us from our Cro-Magnon ancestors.*

SPACESHIPS OF THE MIND

NIGEL CALDER

PENGUIN BOOKS

Author's note

This book is the outcome of encounters with experts who freely shared their anticipations with me. I am entirely in their debt – not only of those whose ideas are quoted in the pages that follow but of the many others who, in their enthusiasm or scepticism, helped to put everything in perspective. Thanks are also due to BBC TV and the co-producers of the associated television programmes, for their support of the project.

Billion is used to mean thousand million.

The television series, *Spaceships of the Mind*, was produced by Dick Gilling and transmitted on BBC2 in 1978. The BBC made the series, in three parts, as a co-production with OECA, Toronto. Alan Stevens and John Baker were the principal film cameramen and the film editor was Christopher Woolley. Visual effects were by Mat Irvine and graphic design by Alan Jeapes. The programmes were presented by Nigel Calder.

Penguin Books Ltd, Harmondsworth,
Middlesex, England
Penguin Books, 625 Madison Avenue,
New York, New York 10022, U.S.A.
Penguin Books Australia Ltd, Ringwood,
Victoria, Australia
Penguin Books Canada Limited, 2801 John Street,
Markham, Ontario, Canada L3R 1B4
Penguin Books (N.Z.) Ltd, 182–190 Wairau Road,
Auckland 10, New Zealand

First published in the United States of America
by The Viking Press 1978
Published in Penguin Books 1979

Copyright © Nigel Calder, 1978
All rights reserved

ISBN 0 14 00.5231 3

Printed in the United States of America by
The Murray Printing Company,
Westford, Massachusetts
Color printed by Rae Publishing Co., Inc.,
Cedar Grove, New Jersey
Set in Caledonia

Picture research by Diana Souhami. Acknowledgement is due to the following for permission to reproduce illustrations. Colour plates are in **bold**. Page **2** California Institute of Technology and Carnegie Institute of Washington. 6 NASA. 9 (top) UKAEA Culham Laboratory, (bottom) Sacramento Peak Observatory/Air Force Cambridge Research Laboratories. 11 NASA. 12 Royal Observatory, Edinburgh. 16 (left) Anita Rimel, (right) Mat Irvine. 19 (top) NASA. 21 Novosti. 23 (top) Alan W. Richards. 24 (top) Oxford Scientific Films Ltd, (bottom) L. B. Lazzopina. **26** (top) Mat Irvine, (bottom) NASA. **30** NASA. 32 NASA. 33 NASA/Keystone. 34 H. Kolm, Massachusetts Institute of Technology. 37 Mat Irvine. 38 (bottom right) Mat Irvine. 39 (bottom) Mat Irvine. 45 (bottom) Aerofilms. 47 NASA. 48 NASA. 49 NASA. 54 Bill Wasson/Photo Services. **57** (top right) Japan Architect, (top left and bottom) Kiyonori Kikutake. **58** Glasshouse Crops Research Institute. 62 New Alchemists. 64 (top) University of Arizona, (bottom) Mat Irvine. 67 (right) Christopher Morrow. 69 (top) John S. Shelton, (bottom) NASA. 72 (top) NASA, (bottom) California Institute of Technology. 75 (top left) NASA, (**right**) Mat Irvine. 78 NASA, (bottom right) Mat Irvine. 83 NASA. **85** (top) NASA, (bottom) Mat Irvine. **88** (top) Ian Scoones, (bottom) New Mexico State University. 90 (top) Novosti, (bottom) NASA. 92 (left) Mat Irvine, (right) NASA. 93 NASA. 97 NASA. 98 (top) Mat Irvine. 100 (top) Sandia Laboratory. 102 (top) Colin Renfrew, (bottom) DOE. 105 NASA. 106 Colin Renfrew. 107 Colin Renfrew. 109 NASA. 110 Novosti. 112 NASA. 115 Mat Irvine. 116 European Space Agency. 118 Ivan Pintar, Cosanti Foundation. 122 (top) NASA. 124 NASA. 132 (top) Cornell University, (bottom) Mat Irvine. 135 (top) Yerkes Observatory University of Chicago, (bottom) Mat Irvine. Diagrams on pages 19, 23, 41, 76, **86-87**, 94, 98, 100, 107 and 130, by Diagram.

The illustrations credited to Mat Irvine are of models made by him for BBC television.

CONTENTS

1

ON BIG IDEAS

When Christopher Columbus was hawking through the courts of Europe his scheme for sailing westwards, alchemists too were seeking royal sponsorship, for research into gold-making. Around the same time Leonardo was sketching a man-powered aircraft, Savonarola was predicting the imminent downfall of civilisation as we know it, and the Scots were inventing Scotch whisky. No one could tell which of these people were justified in their expectations. The only one among them who was demonstrably mistaken was Columbus, because his reckoning of the transatlantic distance from Europe to Asia was a blatant fudge.

The human world, past and future, is shaped by constant pressure from the imagination and ambition of individuals who have big ideas. By a big idea I mean one that will prevail not by decree or even by persuasion but

Enthusiasm moves ordinary people to do extraordinary things, like this astronaut who lived weightlessly for two months aboard Skylab. A resolute minority of our species can open space to human habitation.

because it captures the enthusiasm of people who will struggle against great difficulties to make it happen. Often it will be something that can start in a small way, among a few people; sometimes it will need a measure of public support, as Columbus' did. The big idea must be practicable in the long run or it simply withers, but at the outset its feasibility may be much in doubt even among its progenitors.

Apart from transoceanic navigation, ideas from the past that proved big enough to affect all our lives include animal domestication, crop sowing, metal smelting, organised warfare, organised religion, money, cities, writing, machinery, drama, democracy, organised science, heat engines, football, mass-production, automobiles, birth control, flying, social security, hearing and seeing at a distance, tourism, and organised crime. As some of these examples show, big ideas are not necessarily kindly ones. The test is rather that, by a sort of natural selection, they have survived and exerted a great influence. Take them away and the world would be unrecognisable – yet history or a modicum of prehistorical conjecture allows us in every case to see individual pioneers nurturing the big idea because it appealed to them. The individuals persevered with it, without worrying too much about public opinion, until it became first fashionable, and then the 'natural' thing to do.

This book reports some big ideas about the future of mankind prevalent in the late 1970s, with just a few explanations and comments of my own. The ideas come from a number of individual scientists, ranging in their expertise from astronomy to anthropology, whom Dick Gilling of BBC TV and I consulted during the making of the television series to which the book is an accompaniment. It is not, and was not intended to be, a comprehensive sampling of notions about the future, most of which were running in pessimistic directions. We wanted instead to test the hypothesis that recent advances in scientific knowledge ought to have stirred constructive thoughts.

So we visited distinguished scientists working at the frontiers of knowledge – chosen individuals who we thought might be willing to launch mental spaceships into the world of the twenty-first century or beyond, and share their ideas with us. Gilling and I were open-minded about what the ideas might be and whether they would fit together, or contradict one another. Our prejudice was only that the old dream of a better existence for mankind, based on science, was not yet extinguished. In practice a clear signal quickly emerged from among the free-ranging chatter: the most radical and hopeful ideas clustered around the implications of spaceflight. Here is how I would sum up the main ideas, as they unfold in this book:

People the oceans, the asteroids and the bright deserts of Earth and space. Nurture the diversity of life in other species and ourselves – even by restoring large parts of the Earth to wilderness, taking giraffes into orbit and letting new species of mankind evolve in space. Win immortality for the human spirit and secure it against mere planetary disasters, by taking over the starfields of the Milky Way. But keep an eye open for other beings who may be there first.

Why science in particular remains a source of big ideas for our civilisation is the theme of this introductory chapter. But already I suspect some readers have raised their eyebrows about the extravagant, not to say hubristic notions of that summary. So perhaps I should say a little more about the nature of big ideas. They are not, in the first instance at least, goals in a prescriptive sense – barely even projects. They are more like invitations which people can accept or decline. An American engineer Arthur Kantrowitz caught the spirit of the big idea with his standard answer for those who asked how soon space might be clearly shown to create new hope for mankind. 'Ten years after you stop laughing,' he would tell them.

Those who offer big ideas cannot figure out their eventual consequences. The man who first lined up a squad of troops was vouchsafed no vision of Hiroshima. The people who first travelled for pleasure would have been surprised to see the tourists tumbling out of a jumbo jet. But successful big ideas eventually transform human behaviour and the use of resources. They also overturn the assumptions of the market place and dislodge the accountant, at least temporarily, from the control of affairs. The sums that people spend on preparations for war and on holidays make little sense in relation to subsistence or prudent investment of resources, yet whole economies now revolve around them. And who would have supposed that generals would be looking to their uranium supplies? Or that snowy mountains and sunny beaches had any value, before the invention of skiing and the seaside holiday? If the return on building a cathedral falls due in the life hereafter, how do you show that in the books?

An astronomer who was conceiving a seemingly extravagant programme for listening out for alien intelligences in the universe remarked to us that it would cost less than a small war. An engineer working on a method of shooting lunar rock into space observed that the whole US space programme so far cost less than American women spent on cosmetics in a single year.

Promethean *projects for putting the fire of the Sun in a bottle and achieving controlled nuclear fusion on the Earth throw up curious machines like TORSO (right) at the Culham Laboratory in Britain. If the efforts succeed, they promise virtually unlimited energy for our species. But twenty years'*

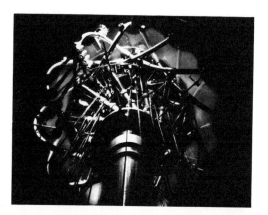

work has gone into learning how to heat and control the fuel in the form of plasma, or ionised gas, at temperatures hotter than the core of the Sun. The eruptive contortions of plasma in the much cooler outer layers of the Sun are illustrated by the solar flare (below) photographed by hydrogen light.

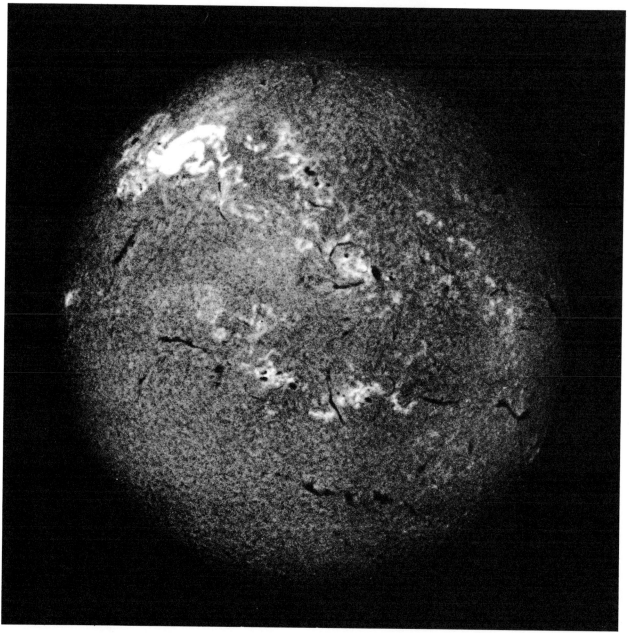

Comparisons of that kind are entirely fair and you can never dismiss a project as 'hopelessly expensive' provided it does not actually exceed the available or foreseeable resources of materials and manpower. Everything hinges on what governments and ordinary folk, after exposure to the ideas, decide to do with their spare resources and their time.

They may well be adventurous. Part of the malaise of twentieth-century civilisation has been a strong tendency to confuse people with livestock. Cautious administrators seem to regard us as beings to be fed and cared for medically, not to be disturbed unnecessarily, and on no account to be put deliberately at risk, except of course in time of war. This attitude has persisted despite all the evidence from psychology of the potentially disastrous consequences of boredom, and despite the way many individuals voluntarily seek action and danger in such pursuits as mountain-climbing and hang-gliding. For a visionary physicist, the late J. D. Bernal, daring was the essence of life. And the astronomer Frank Drake offered us this comment:

'There's no use keeping human beings healthy if there is nothing adventurous, nothing fun for them to do. Well, space provides us with an endless frontier: an endless supply of new places to explore, new adventures, new things we have never seen before, new sources of joy, perhaps even new sources of fear.'

Just what resources count as spare and who has the right to decide how to use them will always be central issues of politics. Rich people and dictators have a greater say than poor people or democratic leaders. Yet the poor seem to have a special desire for cathedrals and a democratic leader can decide to send men to the Moon. Some people, including myself, would argue that the desperate poor of the world have the first claim on any spare resources, but even fair trading with them seems hard to achieve, never mind the big idea of a welfare world. Others, not including myself, say that there are no spare resources: that the human species is living beyond its means and material growth must cease or even go into reverse. Issues like that will continue to be debated, and rightly so. But big ideas enlarge the area of political discourse and often reveal resources that were previously overlooked.

The biggest idea of all is that big ideas are possible: that individual human beings, by taking thought, can change the world. It is not a coincidence that Western civilisation has encouraged that idea most strongly, particularly in the liberal protestant countries of north-west Europe and North America, and has also consistently outstripped civilisations with more fatalistic and illiberal beliefs. And the highest expression of the notion that ideas can alter the world is science. An essential principle of science is that young people are not merely free to disagree with their elders and received opinions but can systematically prove them wrong by calling upon the evidence of nature. Countries where religious, political or cultural inhibitions prevent that process of organised disrespect find it hard to produce original science. Even though science is growing all over the world, Western civilisation still monopolises the major discoveries. It may also be readiest to respond to them. While not unmindful of the Russian imagination and effort in spaceflight, I am reporting the ideas of Western scientists, addressed in the first instance to their fellow Westerners.

There is a scent of renaissance in the air. I think it will turn out that only for a decade, from the mid-1960s to the mid-1970s, did Western civilisation experience momentary doubts about its beacon of hope being science. There was a reaction against *laissez-faire* technology, and science caught the backwash. A salutary if claustrophobic panic set in, as people heeded for the first time old warnings about population and natural resources. It was a bleak period when the big ideas were hijacking, kidnapping and bugging. But at the very time when an epidemic of despair was sweeping through the Sunday supplements, fundamental science was bounding ahead in the most remarkable period of discovery it ever experienced.

Big ideas start with a change in one's perception of the world and the special aptitude of scientists as generators of ideas comes from the way science is forever changing or renewing that perception. In the past it destroyed the cosmology in which the Earth was the centre of the universe and human beings were the specially created masters of the planet; it also revealed the secret of life to be nothing more nor less than a molecular code. The surge of science during the 1960s

Kilroy was here: *a big idea was fulfilled and the track of a handcart marked a human presence in the dust of the Moon. The* Apollo 14 *landing vehicle stood in the distance. The lunar landings showed that men could survive and work on other worlds and the chemical analyses of the rocks they brought back from the Moon encouraged another big idea: namely that large-scale engineering can be undertaken beyond the Earth, using raw materials gathered in space itself. Schemes exist for mining the loose soil of the moon.*

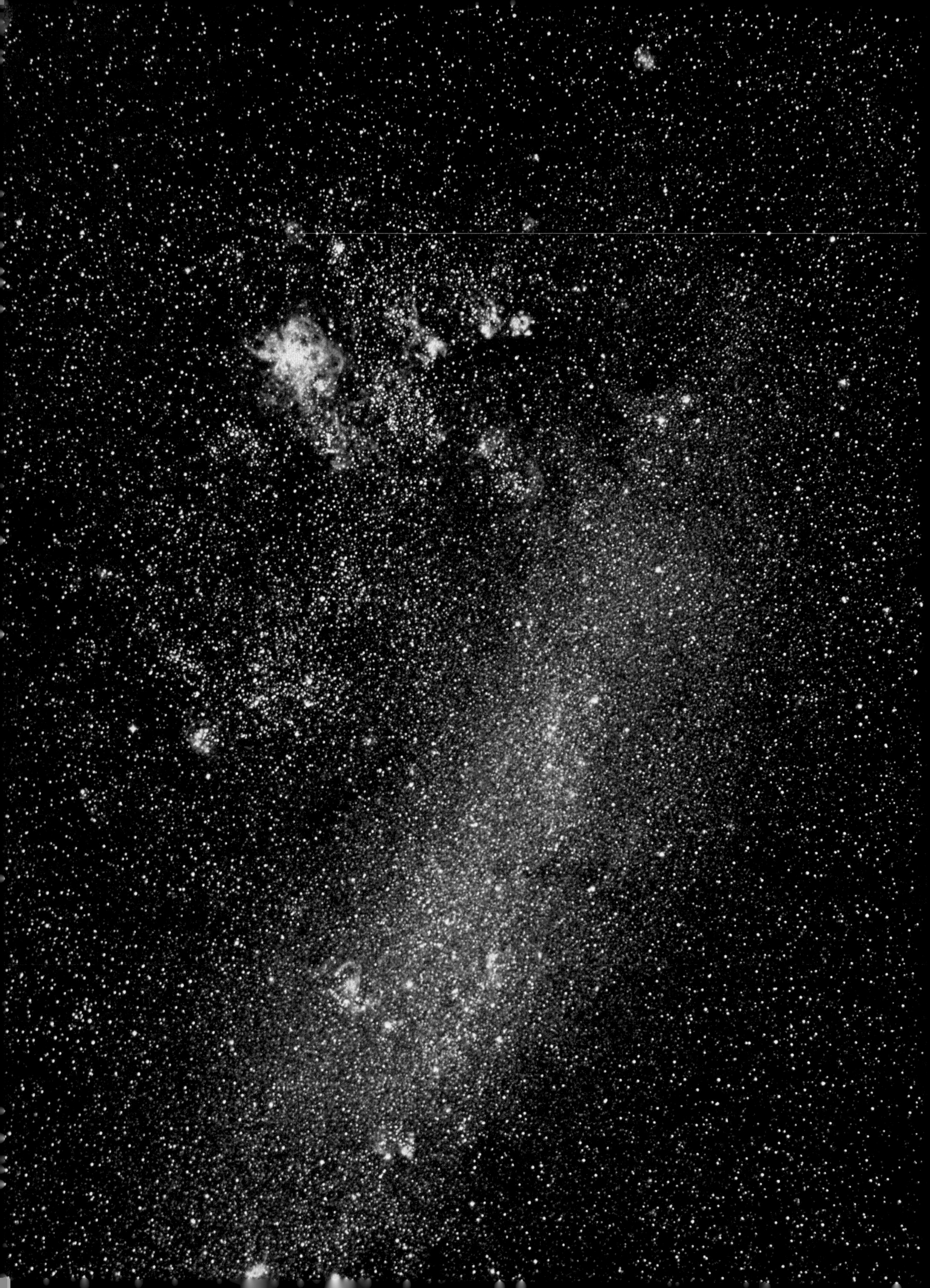

and 1970s brought further changes in perception. The Earth, and its mountains and earthquakes, altered their complexion completely with the confirmation of continental drift and the rise of plate tectonics. The cause of ice ages was found at last – with the uncomfortable rider that the next ice age is not far off. Evolution theory was rewritten in terms of the molecules of life, and there was fresh understanding of how the brain works. A belated reconciliation of the biological, psychological and sociological views of mankind began to create a unified science of human behaviour, making sense of our virtues and vices and telling what kind of creatures we really are. And astronomers found out rather precisely where they stood in space and time, while physicists laid bare general principles underlying all action in the universe.

Astronomers and physicists surveyed billions of light-years of space beyond the confines of the Solar System, and studied the behaviour of the smallest particles of matter under extreme concentrations of energy, thousands of times more intense than anything that occurred even in the heart of the Sun. They came upon many surprises – the explosions of quasars and radio galaxies, the collapse of stars into pulsars and possibly black holes, and novel qualities of sub-atomic matter called charm and beauty. But all these discoveries fitted into a well-understood general scheme invoking a small set of sub-atomic particles and a small set of forces acting between them.

They reconfirmed that, except for its oddest and most violent patches, the entire universe consisted of familiar ingredients of matter shaped by familiar cosmic forces. The chances of finding any new and mysterious ingredients and forces in ordinary matter or living things became slender. And the transparency of cosmic space meant that unlike Columbus, who could not peer beyond the Earth's horizon, one need expect no hidden continents in the nearer regions at least. Accordingly one could take stock ever more confidently of the chemistry of the universe.

A vivid summary of this knowledge, embedding human beings firmly in the universe of stars and galaxies, came from one of the scientists whom Dick Gilling and I went to see. He was John Lewis, a cosmic chemist at the Massachusetts Institute of Technology. A comprehensive theory of the composition of the planets and moons of the Solar System, which Lewis brought out in 1971, had become a guide for interplanetary explorations. But he also embedded human beings in the universe of stars, galaxies and planets, saying to us:

'We begin with a universe of hydrogen and helium, in Big Bang cosmology, and the most abundant atom in the human body is hydrogen, the raw material of the universe. After that, the fusion of nuclei in stellar interiors produces oxygen and carbon – these are the next two most abundant elements of the human body. The heavy elements are produced in later stages of stellar evolution. And all of these materials are then spread around in interstellar space by the explosions of stars. It is a necessary precondition to the existence of human life on Earth that there have been already several cycles of stellar evolution . . . Our planet is the result of one stage of planet building which is a substep of one stage of stellar formation.'

Lewis went on to describe the geological cycles in which all life on Earth is enmeshed. For example, materials from living things find their way into sediments on the ocean floor, which creeps along at a few centimetres a year before being shoved down into the Earth's interior at the edge of a continent. There the high temperatures throw the material up and out through volcanoes, to become available again for plants and animals. Lewis drew a moral from all this. 'The unity of nature must be reflected by the unity of our appreciation of it.' He wanted to reconcile our intuitive feelings about nature with the intellectual analysis of it. By this, as a devotee of transcendental meditation, Lewis meant integrating consciousness and objectivity – an aim he thought more important than manipulation of the universe. But my concern here is with the integration of mankind and the cosmos in the objective ways.

In talking with scientists about the human niche within the vast and ancient universe revealed by modern astronomy, I detected very different kinds of 'intuitive feelings about nature'. Some saw us diminished: all we could do was try to snatch a little dignity in cultivating our planet, and draw a little pride and rationality from our understanding of the universe. Another feeling, reflected strongly in this book, was that human beings had the collective knowledge and skill to start transforming the universe to their own purposes.

Every speck of starlight is a source of energy like the Sun and, in principle at least, it is a potential powerhouse for abundant life. The Large Magellanic Cloud, a nearby galaxy and a satellite of our own Milky Way, may be forever beyond human reach. Yet it shows the capacity of the forces of the universe to create luminous wealth almost without limit. Some scientists find the scale daunting; to others it is an invitation.

For those preferring the bolder view, the knowledge of cosmic chemistry and forces became a map of available resources and the means of handling them. You could make a fairly precise inventory of all the materials on Earth and list all possible means of converting energy. That gave an envelope of possible human action, much less restrictive than current economics might suggest. Then, with lower precision perhaps but by the same principles, you could do the same for space, and discover it to be a source rather than a sink of wealth.

I have gone back to the Big Bang and its sequels to help explain why big ideas flow from fundamental discoveries in science, sooner than by extension of familiar technology. It was not the officials of government space agencies who suggested that ordinary people might go and live in space, but physicists who spent their time finding out how the universe was built. They possessed a sharper sense of the springs of action in nature; also the higher numeracy which enabled them to reckon the nickel content of an asteroid at one moment and at the next to estimate the sunlight needed to feed a million people. Every day their research liberated their minds from mundane prejudices about what was 'natural' and 'practical'.

The effort to steal the fire of the Sun and put it in a bottle was a product of that kind of thinking. Whether one was looking ahead to the need to cope with the next ice age or simply reviewing energy policy for the end of the twentieth century, the hoped-for prize of thermonuclear fusion would transform the prospects. Once tapped, the prime energy source of the universe would be essentially limitless. But that objective could never have been stated without the fundamental discoveries about natural nuclear reactors.

Less than a quarter of a century passed between finding out that the fire of the Sun and the stars came from the fusion of the nuclei of light elements, especially of hydrogen to make helium, and its terrible imitation on Earth in the form of H-bombs. The natural process required very high temperatures and long periods of time. The time could be shortened to an instant by using nuclei a little heavier than ordinary hydrogen, but attaining the high temperatures remained the problem. You could do it by exploding an A-bomb, so igniting an H-bomb, but when people tried to heat the fusion fuels in milder ways for peaceful purposes the hot gases wriggled and squirmed out of their magnetic bottles. It was a messy business, but researchers in the USSR, the USA and Europe made gradual progress. By the 1970s the consensus among them was that practical thermonuclear reactors might be operating in the 1990s.

There were several promising ways of doing it, by magnetic confinement, by heating pellets of fuel with lasers or electron beams, and so on.

The first generation of fusion reactors will almost certainly use deuterium (heavy hydrogen) and lithium as fuel. Obtainable from sea-water, with lithium as the limiting factor, these will supply mankind with energy at twentieth-century rates for 100 million years. The snag will be the outpouring of potentially harmful neutrons from the reactor. But just the same kind of reasoning that encouraged the attempts to imitate the energy of the Sun led some fusion specialists also to see that, if they went to the planet Jupiter, they could fetch back a thermonuclear fuel much less troublesome in that respect. Moreover, fusion rockets could be the way of getting people to Jupiter – and later to the stars.

In what follows, I shall avoid adopting any particular scenario for future operations in space. Instead, the book tells of various schemes of knowledgeable enthusiasts. They span a wide range of motives and methods for using the resources out there. Every proposal has its prizes and its difficulties; in each case the difficulties are almost certainly greater than the protagonists have yet realised. And there are competing ideas for making a richer life on the Earth, which I shall not neglect. But if humanity cannot support itself without technology, it may be well advised to take some of its technology off the Earth.

Listening to the experts, I have been persuaded that the difficulties will be overcome. One reason is the large choice of technical, economic and political possibilities: *something* will work. Another is that the physics and chemistry are not merely sound but compelling: the unused energy and material of the Solar System is enormous. Getting the biology right will be tricky, as we shall see, but not insuperably so.

The third reason for believing that the movement into space will occur is the hot breath of enthusiasm itself. The individuals expressing it comprise a minority of our species, but the recruiters would have no difficulty in manning a space settlement today. Whether or not the reader and the author share the enthusiasm is of little consequence: others have it. I suspect that life in space will often be like going around the world in an early sailing ship, where human beings endured pigsty-like conditions cut off from friends and families, and where seasickness was curable by death from thirst, scurvy or drowning. The most amazing aspect of the European break-out across the oceans was not the skill of the famous captains but the doggedness of anonymous seamen who survived the voyages and signed on again.

They and their passengers were a small fraction of the population of Europe, but they built empires and created new nations. The same sort of process will begin again and will alter human circumstances even more profoundly.

I do not think living in space is a nice idea. Apart from the danger and discomfort, people will no doubt fight and exploit one another out there, and our descendants will evolve into weird creatures that we might hardly acknowledge as human. But the proposed break-out into the universe is *exciting* and it has the hallmarks of a successful big idea – of something that people will actually do, almost regardless of the opinions of bystanders. Dreams of adventure, riches or freedom will pull them, while the pressure of population pushes them from behind.

The idea that the resources of the universe might be transformed for the purposes of life does not lie far from the main stream of twentieth-century science. Proof came in the entirely serious contemplation of a habitable universe by astronomers who began the search for signals from intelligent beings living on the planets of other stars. In 1959, two physicists of distinction put this programme on the scientific agenda: Giuseppe Cocconi of the European Organization for Nuclear Research and Philip Morrison, then at Cornell. During the following two decades much thought went into the biology of the putative beings and into the possible means of spotting their radio signals or light signals. Recent studies have considered the detection of a spaceship visiting our parish of the Galaxy.

In my opinion that last thought should bring a blush to every human cheek. Imagine that alien spaceship cruising among the stars and running now into the babble of radio and television signals leaking from the planet Earth. What will the cosmic neighbours think, as burst after feeble burst testifies to our barbarity? Baboons who play Bach – you can almost hear the derision echoing along the Milky Way as the word passes about our quarrelsome species. But perhaps most mystifying for the eavesdroppers will be the economists whose cry is 'scarcity' and who explain so learnedly to the poor why they have to stay poor.

We can picture the aliens pausing to refuel with the gases of Jupiter before moving off. With one casual flick of their motor they will expend in a few days more energy than the human species has used in the whole of its history. Fecklessness might be the main theme of their report on the new-found source of radio pollution inwards of Betelgeuse. It emanates from beings who have mastered a lot of physics, chemistry and biology and yet let their children starve – while all around their planet the energy of the mother star runs to waste in a desert of space.

J. Desmond Bernal *(1901–71) was a crystallographer of high renown and the most distinguished scientist to speak and write emphatically about the possibilities of living in space, long before the Sputniks. He imagined man-made globes inhabited by large numbers of people. Gerard O'Neill and his colleagues conceived, in the 1970s, a large spherical settlement orbiting the Earth with 50,000 people – and they named it the Bernal Sphere (model, right). It departs from Bernal's original idea for weightless life in space by rotating, to simulate gravity with a gentle centrifugal effect. On either side of the central sphere, where people live, are hooped structures that contain the agricultural areas. These enable the settlers to feed themselves. A ring of mirrors guides sunlight into the settlement.*

2

SANTA CLAUS MACHINES

Big ideas have to be simple to state; otherwise they will not penetrate the hubbub of human life to command any sort of attention, or capture the imagination of potential supporters. In this chapter I present ideas from three physicists and a biologist which seem to me to underpin what follows later in the book. They also illustrate how very simple propositions can come out of science, heavily laden with implications. Contrary to a popular misconception science does not become more and more complicated and baffling with the passage of time. The scientist's aim is understanding, which means penetrating the hubbub of nature to find general principles. A sense of cosmic history results – and it provides a useful preamble to the first of the ideas I want to report.

One can now point to natural machines that systematically separate and recombine pieces of cosmic material. To make the Earth the way it is, several such machines had to operate. By an advanced stage in the evolution of the universe, the heat and gravity of the newborn Sun were sorting out the cloud of materials around it. The very abundant lightweight gases were driven from our sector of the Solar System; those which did not escape entirely lighted on the big outer planets. By that process of planetary distillation the Earth was formed as a rocky planet from the heavy material that survived in its sector. Then nature started another machine running. Again it was operated by gravity and heat – partly heat left over from the formation of the Earth, partly heat from radioactive energy inherited from the explosions of stars that had created the elements at an earlier stage. This machine caused molten iron to sink to the core of the Earth, leaving a thick shell of rock, the mantle, floating on it. The planetary cake was topped by a crust rent by volcanoes, which poured out steam to fill the oceans and also the gases that created the soup of chemicals whence life began.

Life was a third sort of machinery for rearranging the atoms of the universe, but now assembling elaborate molecules which could convey and execute instructions about how to stay alive. It ran on sunlight, which enabled the Earth's cloak of green plants to take water and carbon dioxide, and lesser quantities of other nutrients, and produce living tissue and oxygen. Some dead plants made coal and oil, which helped to leave oxygen free in the atmosphere. Animals fed on the plants and breathed the oxygen. And nature's living machine eventually fashioned people – along with the deer, salmon, wild honey, nuts and all the other gifts that human beings enjoyed freely until proto-economists taught them to know better.

Knowing this history, my imagination was ready to be captured by a man-made extension of those natural processes for the rearrangement of materials: the Santa Claus Machine. Besides being a big idea in its own right, it symbolised the general possibilities that will flow from the human manipulation of the chemical ingredients of the universe. Telling us about it was Theodore Taylor, a man steeped in the faustian technology of the twentieth century. He was formerly involved in the design and testing of American nuclear weapons; latterly he became an outspoken advocate of arms control and an enthusiast for small-scale energy production on Earth using sunlight. In 1977 Taylor gave up an engineering professorship at Princeton to go to Washington, D.C. to run the International Research and Technology Corporation.

Among his many interests was the challenge of exploiting the raw materials of the Solar System. Taylor described the system that he had in mind, like this:

'It's possible to imagine a machine that could scoop up material – rocks from the Moon or rocks from asteroids – process them inside and produce just about any product: washing machines or teacups or automobiles or starships . . . Once such a machine exists it could gather sunlight and materials that it's sitting on, and produce on call whatever product anybody wants to name, as long as somebody knows how to make it and those instructions can be given to the machine. I think the name Santa Claus Machine for such a device is appropriate.'

As visualised by Taylor, the machine will operate automatically, without any immediate involvement of human beings. Its central principle will be the sorting of the raw material into all the individual chemical elements which it contains. That will be done by a giant version of the mass spectrograph – an analytical instrument from the physics laboratory which converts material into a beam of ionised (electrified) atoms travelling in a vacuum. It then deflects the beam with a magnetic field. Because lightweight atoms accelerate and swerve more readily than heavy atoms the various elements and isotopes in the beam can be sorted, atom by atom.

In space, nature will provide a ready-made vacuum, allowing the mass spectrograph to be scaled up into a large refining plant. But it will need big magnets. In Taylor's scheme these may be superconducting electromagnets, working at low temperatures and offering no resistance to the flow of electric current; alternatively, a large machine near the Earth may exploit the weak but extensive magnetic field of the planet itself. At reasonable distances from the Sun, and certainly in the vicinity of the Earth and the Moon, ample power for the Santa Claus Machine will come from solar energy. In the windless vacuum of space one can build large yet flimsy mirrors, like gleaming parachutes. They will focus enough sunshine to vaporise rock.

To live up to its name, the Santa Claus Machine must make the stocking-fillers, from the stockpiles of very pure materials created by the mass spectrograph. The materials can be recombined or mixed to make any compound or alloy. According to Taylor, the manufacturing processes will be quite different from what one sees going on in a steel mill or car factory on the Earth. They will take full advantage of the vacuum and weightlessness of space – for example, making parts simply by revaporising the selected materials and depositing them on moulds. Given a suitable range of automatic tools and process controls in the system, people will simply have to ask for what they want, and tell the machine how to make it.

Santa Claus Machines in space will supply raw materials and manufactured goods to the Earth. In the

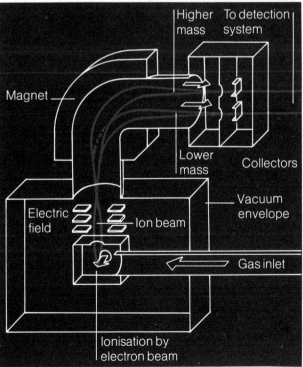

Santa Claus Machines *would take raw materials available in space, for example on the Moon's surface shown in this Apollo photograph. They would separate the materials into their elements, atom by atom, using a mass spectrometer. The diagram shows the principle of a laboratory mass spectrometer (VG Micromass 602D) which separates electrically-charged molecules of two different molecular weights. The magnet deflects the lighter molecules more sharply. In the natural vacuum of space the mass spectrometer could be scaled up and elaborated greatly. The Santa Claus Machines would then be programmed to use the separated materials in the automatic manufacture of any desired products.*

beginning, when operations in space are still costly, the products will have to be ones that are very expensive on Earth, in price per pound, and yet command very large markets. In Taylor's opinion, one may have to talk of markets of hundreds of millions, or even billions, of dollars a year. The most attractive products will be materials like aluminium and titanium which require a lot of energy for their separation. Some of the lunar rocks are far richer in titanium than the titanium ores on Earth. In the long run, practically any material, pure or mixed, will be cheaper and easier to make in space using extra-terrestrial sources. For Taylor, the most appealing benefit of large-scale production in space will be its disconnection from the Earth's biosphere, which will be relieved of pollution.

Apart from serving the Earth's inhabitants Santa Claus Machines will have a much wider role, in developing the resources of the Solar System. They can orbit the Earth, the Moon or the Sun or latch on to asteroids, the minor planets. Re-adapted to conditions of gravity and wind, they will be able to sit on planets or the moons of planets and gradually transform them. Their manufactured products can include space settlements for human habitation – and even new Santa Claus Machines. Taylor's proposal for total separation of the elements means that all of them will come out of the same melting pot, in proportion to the amounts present. Potentially more precious than gold and platinum will be the extraction from extra-terrestrial material of the elements indispensable to life – hydrogen, carbon, nitrogen, oxygen, and so on.

Out among the distant planets, where sunlight becomes feeble, the Santa Claus Machines will run by nuclear power. The Soviet satellite, *Cosmos 954*, falling over Canada in 1978 with a reactor on board, prompted more cautious policies about using nuclear sources of energy in space, near to the Earth. Nevertheless, part of Taylor's vision is that nuclear power will come into its own in space, avoiding or curing some of the problems associated with nuclear power on Earth. Nuclear fuels, for example, will be prepared far beyond the reach of would-be nuclear terrorists on Earth, and the best way of getting rid of radioactive waste material may be to put it into the Sun, or else to shoot it right out of the Solar System.

But not dumping it too casually into space. At a London dinner party in 1977 a lady was talking in high excitement about a news item that day telling of a suggestion for disposing of nuclear wastes in space; also of protests against it, the objection being that space should not be contaminated because people were going

to live there. 'That's just what Bernal always said,' she declared: 'that people would live in space.' The lady was the widow of the physicist J. Desmond Bernal. I was glad to be able to tell her that scientists were pursuing his idea of constructing permanent settlements in space; and that a prime configuration under study in the USA was called the Bernal Sphere.

Bernal was a scientific polymath, brimming over with ideas on every subject. 'Nature needs to be attacked from many and unexpected sides,' he liked to say. I recall one conversation with him which ranged from house-building to ice to continental drift to the origin of life: on each topic he had published original contributions. More especially he was one of the founders of the modern science of X-ray crystallography; he also made the first X-ray analyses of biochemical crystals and viruses, opening the road to molecular biology. During the Second World War, in Combined Operations, Bernal worked on the artificial Mulberry Harbour for the Normandy landings, and on an unsinkable aircraft carrier to be made of ice.

Although Bernal was obsessed about the role of the scientist in society, his unrepentant communism denied him the chance fully to play that role in Britain. But his legacy to the world included a short book that he published in 1929, while a young lecturer at Cambridge. It was called *The World, the Flesh and the Devil* and it anticipated to a humbling degree the broad topics about the future of mankind that were to become fashionable half a century later: the colonisation of space, the restoration of the Earth to a more natural state, the self-directed evolution of mankind – as well as many detailed suggestions from solar sailing to interconnecting brains. Bernal saw much of his scientific life as an elaboration of the seeds of ideas sown in that book.

By Bernal's prophecy, humanity will emancipate itself from the Earth and the majority of people will live in man-made globes orbiting about the Sun. In this vision Bernal was anticipated to some extent by writers of science fiction, and notably by Konstantin Tsiolkovsky in Russia who, in 1903, described a space habitat that would rotate to create artificial gravity by centrifugal force. Bernal, on the other hand, opted for weightlessness, with all its conveniences and inconveniences.

His reasoning started with the small size of the Earth in relation to the Sun. It intercepts less than a billionth part of the supply of radiant energy pouring out from the Sun. That is a trivial calculation from schoolroom physics, yet only a few people like Bernal grasped its cardinal meaning for the future of our species. When we

A Soviet model *of a space station, at the Cosmonautics Museum in Kaluga, celebrates the ideas of Konstantin Tsiolkovsky (1857–1935). The photograph below shows him at the age of 70. Tsiolkovsky was a schoolteacher in Kaluga, and a lifelong victim of deafness. He envisaged liquid-fuelled rockets and spinning space stations.*

learn to live on this wasted solar energy, Bernal perceived, 'the possibilities of the spread of mankind will be multiplied accordingly'. According to his prediction the globes will be constructed in the first instance by hollowing out small asteroids. Some 20,000 or 30,000 people will inhabit a sphere having a diameter of about ten miles and possessing vast membranous wings to increase its capacity to gather sunlight.

After Bernal's globes came the 'Dyson Sphere'. Before Bernal died in 1971 another distinguished physicist, Freeman Dyson, was already addressing similar themes with equal vigour. Born in England, Dyson settled in the US, becoming professor at Cornell and then at the Institute for Advanced Study in Princeton. He worked in the mainstream of the postwar physics of particles and forces and it was in that connection that I first met him, to learn from him about the delicate notion of broken symmetry. But Dyson seemed appropriately the first person to go back to when I was contemplating the present exploration of big ideas about the future. I knew he had been caught up in the US government's Project Orion, which studied the possibility of a manned space rocket to be propelled by nuclear bombs; also that he had spun many Bernal-like thoughts about long-term possibilities for mankind. They included 'the greening of the Galaxy' and the extraordinary concept that came to be called the Dyson Sphere – the idea of putting a box around the Sun.

The discussions among physicists and astronomers about detecting alien intelligences on the planets of other stars prompted Dyson to think what technically advanced species might do to alter their surroundings on a large scale, in such a way that the consequences could be seen from the Earth. In 1959 he published his 'reasonable expectation': that an intelligent species with an expanding population, living in a planetary system like the Solar System, will eventually rearrange the material of the planets in order to create, in effect, a thin spherical shell around the parent star.

It will not be a single rigid sphere by any means, but will consist of a large number of artificial space cities each pursuing an independent orbit about the star. The structures in this 'artificial biosphere', lying at a comfortable distance from the star, will together exploit virtually all of the available radiant energy. That will tend to black out the visible light of the star, from the view-point of an outside observer. The swarm of objects will though, by virtue of their warmth, emit an equivalent amount of infrared radiation. Accordingly Dyson proposed looking for 'infrared stars' as possible sites of advanced civilisations.

It was part of Dyson's reasoning that human beings could, if they wanted, accomplish such a feat within a few thousand years. If our own species attempts to enclose the Sun with a Dyson Sphere it will be a vast undertaking. It might for example mean dismantling an entire planet. But mankind's population and industry would be correspondingly large – perhaps a million million times greater than in the twentieth century.

Building the Dyson Sphere around the Sun began, in a modest way, with the launching of artificial satellites and space probes equipped with solar collecting panels as the source of power. They supply us with information rather than support for life, but they do intercept a little of the sunlight that otherwise runs to waste. As for creating a habitable Dyson Sphere, Dyson's own opinion is that it will not start with a big architectural design but will grow by gradual agglomeration. As he described it to us:

'We will go into space, as I hope, in large or small groups in different kinds of structures. Some of us will colonise asteroids, some of us will colonise planets, some of us may colonise comets. And so we will gradually distribute ourselves in more and more places . . . One has to envisage some kinds of traffic laws – streaming of artificial space cities with orbits coordinated so they don't bump into each other. And so, after you've been doing this for a few thousand years, you will have built up some kind of organised design in which these artificial structures are more or less filling up the space.'

The rearrangement of the material of the Solar System will be, it seems to me, a plausible enough extension of human activity on Earth. Bertrand Russell once defined work as either moving pieces of the Earth's surface around or else telling other people to do it. More biologically one can say that the whole point of being an animal capable of independent movement is to be able to gather resources that are otherwise beyond reach – resources of food and nest-building material created by the energy of sunlight falling on some other patch of the Earth's surface.

Scampering about more than most animals did,

Freeman Dyson *is a theoretical physicist and the boldest expounder of ideas for making the inanimate universe support life. His notion of an artificial biosphere around a star has come to be called the Dyson Sphere. An advanced civilisation creates such a swarm of orbiting platforms that eventually they intercept nearly all the radiant energy of the parent star. The dense orderly swarm looks like a heavy veil.*

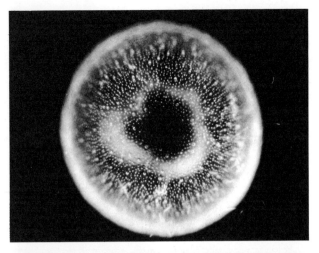

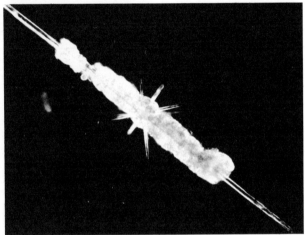

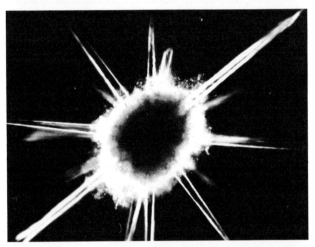

Not spaceships *but microscopic marine animals called radiolaria that float by the billion in the oceans. Amazingly, they build their varied and intricate frames from materials scarce in sea water – silica or strontium sulphate. The radiolaria, like the millions of other living species, are evidence of the extraordinary capacity of organisms to evolve to exploit the various ecological niches on the Earth – often in improbable ways. It is almost as if the DNA, the hereditary material on the genes, were driving organisms to use every possible source of energy and raw materials to ensure its own replication. Sol Spiegelman (right), a distinguished molecular biologist and cancer researcher, likes to say that DNA seems to have invented man to explore the possibility of extra-terrestrial replication.*

human beings were at first content to harvest the gifts of nature. But then they decided that nature's arrangements were less than perfect. They took seeds from one place and sowed them in others. In due course they began shifting materials around the Earth on a grand scale, shipping nitrates from Chile, oil from the Persian Gulf, wheat from North America, iron ore from Australia, and so on. When human beings go in search of energy and materials beyond the Earth – aided no doubt by Santa Claus Machines – they will be continuing the quest of the ancestral worms, which did not wait for food to drop into their mouths but burrowed into the mud of the seabed in search of it.

Sol Spiegelman formulated this kind of idea in terms of the genetic messages of life. We consulted him as an outstanding molecular biologist who thought about the broad sweep of life on Earth. Spiegelman made headlines in 1965, while at the University of Illinois, by becoming the first scientist to produce a living entity – an active virus – from non-living material. He went on to Columbia University in New York to carry out cancer research. But he found time also to organise some remarkable experiments demonstrating that inanimate nucleic-acid molecules could evolve in a test-tube by the same principle of natural selection as operates among living organisms. That work shed light on the origin of life on Earth, when nucleic acids were the first entities with the capacity for self-reproduction. They were the ancestors of the genes, nowadays generally incorporated in molecules of DNA.

Only half-jokingly, Spiegelman put mankind into a molecular perspective, from the viewpoint of the genes. He said that the nucleic-acid molecules began to organise an environment around them that would optimise the generation of energy for their replication – in other words, they created living cells. And Spiegelman's story continued:

'Then the cells got together to form the simple and higher animals and plants. And these were obviously designed to permit nucleic acid to explore the various ecological niches on Earth, like the water, the air and the earth. It went on for a long time, until virtually all ecological niches on Earth were occupied. Then at some point DNA invented man. And for a long time we really didn't understand why – until a few years ago it was clear that DNA invented man to explore the possibility of extra-terrestrial life, as another place to replicate.'

Keeping his tongue in his cheek, Spiegelman suggested that the genes were very careful in devising man.

They left him not quite smart enough to make an ideal existence on this planet, but able to pollute it sufficiently so that there would be pressure on him to look for other places to live. That, of course, would serve the purpose of DNA perfectly. When we pressed him, Spiegelman did not of course think there was any superplan which pre-ordained our exploration of the universe. Indeed in social terms he considered it to be something of a cop-out, an attempt to run away from the problems on the Earth. Yet he thought the biological and intellectual pressures were probably sufficient to make it happen anyway.

There is nothing mysterious about the biological pressures of which Spiegelman spoke – only the awesome arithmetic of human numbers. As Dyson pointed out, to multiply the present population a million million times will take only about 3000 years at a growth-rate of one per cent per year. That is half the growth-rate of the mid-twentieth century. Bernal indicated the wasted sunlight and raw materials of the Solar System, which can indeed make life incomparably more abundant, but they will not sustain an indefinite growth of population. The Dyson Sphere defines a limit, when the whole of the Sun's radiant energy is being used to support human life: then Malthus may have his due. Even if non-solar sources of energy help to maintain life the benefits will be marginal, given such an enormous population. The export of people to settle near other stars will bring significant relief only to the interstellar migrants themselves.

Our species satisfies all too easily the reproductive requirements for peopling the universe. I hope we shall not let that become the compulsion. It will be one thing to go freely into space, to plant life in the cosmic deserts and so to gain elbow room; quite another to recreate Calcutta across the entire sky. That is my own reason for thinking that the Dyson Sphere must never be completed. But the limit is not a near one. A billionfold increase in the human population will be comparatively easy to support in the environs of the Sun, if that is what we want.

Bernal, Dyson and Spiegelman described the elementary logical basis in physics and biology, for a human break-out from the Earth. Theodore Taylor's Santa Claus Machine defined a rational long-term technology. But another scientist was thinking in more detail about how the process could begin in the twentieth century – and was even offering it as a remedy for the population problems of the twenty-first century.

Geological prospecting *on the Moon will be a necessary preliminary for mining operations. Astronauts like the geologist Harrison Schmitt of Apollo 17 (below) could survey only small regions. The proposed Prospector spacecraft (model, right) was therefore conceived as a way of scanning the entire surface with instruments capable of identifying chemical elements by the characteristic radiations they emit. It would orbit over the Moon's poles.*

3

In August 1974 a short letter appeared in the London scientific journal *Nature* arguing that the land available for human habitation can grow more rapidly than the human population increases. It could be done by using material from the Moon to build new 'land' in the form of large habitats, orbiting the Earth and accommodating large numbers of people. A graph accompanying the letter showed the human population in space outstripping the population on Earth in the latter half of the twenty-first century. By then the Earth's population would be diminishing because of mass emigration into space. It would level out before AD 2200, perhaps at what it was in AD 1910 – around 1200 million.

That was Gerard O'Neill's first formal publication of his vision of the future of mankind in space. Two American scientific journals had declined to publish his ideas, despite the fact that O'Neill was a professor of physics at a distinguished university, Princeton. But when at last *Nature*, and *Physics Today* in the US very soon after, carried his message, interest blossomed. By 1977 many newspapers and magazines were describing O'Neill's concepts, he had brought out a popular book with the resounding title *The High Frontier*, and he had a large following of experts, students and non-academic enthusiasts. O'Neill and his immediate colleagues were engaged, with the National Aeronautics and Space Administration and other institutions, in studies and conferences to develop their ideas. According to the Princeton prospectus, the first substantial space settlements will be functioning before the end of the twentieth century, and will eventually help to solve a whole catalogue of human problems: over-population, shortages of energy and food, pollution by industry, climatic change and the limits to economic growth on Earth.

O'Neill's big idea differed from ordinary astronautical thinking in suggesting that the structures for human habitation should be built almost entirely in space. To fabricate a space settlement weighing a million tons or more on Earth and then to launch it piece by piece into orbit against the Earth's gravity would be extravagant, to put it mildly. But when one began to appreciate that space in the vicinity of the Earth was rich in suitable raw materials in conditions of less gravity, and moreover possessed an intense and uninterrupted supply of solar energy, the task seemed to become much simpler. It was then a matter of sending construction workers into space, with a minimum of tools, supplies and systems for supporting life, and letting them get busy building the first settlement in orbit. Thereafter, the settlers themselves would build new settlements and the numbers would grow rapidly, doubling and redoubling every few years.

About the older attitudes, O'Neill remarked that people had thought in terms of space as a sterile environment, to be approached in much the same way that Scott or Shackleton went into Antarctica. You would carry everything with you that you needed, and consume it along the way. Then you would leave the sterile environment as quickly as possible. O'Neill, for his part, recognised that the stock of chemical elements available in space and the unceasing input of energy from the Sun provided the essentials for both agriculture and industry, allowing people to fare even better in space than on the surface of the Earth. Indeed, once you began to look at the attractions of space instead of harping on the obvious disadvantages, you saw virtually inexhaustable supplies of energy and materials, efficient transport without friction or drag, and extraordinary possibilities for engineering in the absence of gravity and air.

Tsiolkovsky, Bernal and Dyson were some of the scientific and science-fiction writers who dreamed of man-made settlements in space long before O'Neill devised his versions. But he discovered their writings only after his own calculations were well advanced. More significantly, he was the first to describe engineering methods for settlement-building, to declare it a feasible project for his own generation and to back the idea with enough knowledge, energy and determination to make people pay attention to it. He was at pains to argue that living in space will be agreeable – fun, even – and that the conditions in a space settlement will approximate to conditions on Earth just about as closely as the designers and inhabitants may wish. Rotation will simulate gravity by centrifugal force and artificial hills and rivers will be provided as desired. Describing life beyond the Earth in *The High Frontier*, O'Neill used the device of imaginary letters written by an inhabitant:

'. . . We live in Bernal Alpha, a sphere about five hundred metres in diameter, with a circumference inside at its 'equator' of nearly a mile. We have track races and bicycle races that use the ring pathway . . . Alpha has a Hawaiian climate . . . Our apartment is about the same size as our old house on Earth and it has a garden. Alpha was one of the first habitats to be built, so our trees have had time to grow to a good size . . . Ballet in 1/10 gravity is beautiful to watch: dreamlike, and very graceful . . . There are almost as many different kinds of local government as there are national styles within the colonies: ours happens to be a town-meeting style . . . Fresh vegetable and fruit are in season all the time.'

The general concept is of hollow, spinning space settlements inside which people will live with their feet pointing outwards away from the axis of spin. Looking up, they will see the 'other side' of the settlement overhead. Ingenious systems of mirrors will guide sunlight in past the cosmic ray shields and cause it to fall with suitable intensity on the living areas. The light may not appear sun-shaped, nor will it track across the 'sky': night will be simulated by adjusting the mirrors.

The proposal had just the right provenance to suit our hypothesis about fundamental science being the most promising source of big ideas for the future of mankind. O'Neill was a high-energy particle physicist by trade, one of the people whose activities seemed to outsiders to become ever more remote from practical human concerns, as they interrogated nature ever more deeply about the laws of creation. But if answers seemed esoteric it was partly because nature had tricks richer than common sense was used to, partly because the answers were incomplete and the physicists themselves were still groping. There were illuminating advances in the 1970s, including the discovery of 'charmed' particles at the Stanford Linear Accelerator Center. O'Neill could

Gerard O'Neill (*right*) *and some of his space-settlement associates, including Henry Kolm (second from left) and Eric Drexler (third from left). The NASA painting represents the interior of one of the most roomy space settlements envisaged by O'Neill.*

DON DAVIS

claim indirect credit as the young man who, in the late 1950s, proposed building the first machine at Stanford for storing beams of electrons and colliding them head on. That was the ancestor of the famous SPEAR machine, devised by Burton Richter for colliding electrons and anti-electrons, with which the discovery of charm was made. Long after he began his crusade to put people into space O'Neill remained involved in high-energy experiments running at SPEAR.

By any reckoning, counting K-mesons from electron/anti-electron annihilations was a different game from designing space settlements. But the habits of thought were not dissimilar: conjuring up machinery that no one ever imagined before; the patient teamwork needed to work out an idea using a great range of knowledge and skills; the higher numeracy that gave a quick, intuitive sense of what was physically feasible, whether in estimating the chance of a collision between sub-atomic particles or in computing the strength of an aluminium cable binding a space colony. Above all, the interrogation of nature in high-energy physics called for the continual exercise of bold imagination by the cleverest people of their generation. And from the midst of that community, as a former student (at Cornell) of people like Philip Morrison and Freeman Dyson, came O'Neill. He was an amateur aviator and Project Apollo stoked his imagination; he had himself shortlisted as a potential astronaut. His imagination was still burning as he evolved ideas that bypassed the newly conventional thinking about astronautics.

By the time the astronauts first stepped on to the Moon, in July 1969, disenchantment with science and engineering was widespread on American campuses. It went with the student revolt against university administrators, with horror about the war in Vietnam, and with growing concern about damage to the natural environment. Science and engineering undergraduates found themselves criticised by their fellow students in arts, and social science. Concerned about this state of affairs, O'Neill invited some of the youngsters in his freshman physics course at Princeton to attend an extra weekly seminar and to consider afresh how available science and engineering might improve the condition of mankind.

Was the surface of a planet really the right place for an expanding technological civilisation? Starting with that question, O'Neill gave the students the task of calculating how large a cylinder could be built for space habitation, holding a normal atmosphere and five feet of soil and rotating to simulate gravity. The answer turned out to be several miles in diameter. O'Neill was surprised and soon convinced himself that what had started as an exercise for students was practicable and important. He continued work on the idea in his spare moments, evolving a space-settlement design with two counter-rotating cylinders, each carrying three big mirrors, three windows and three strips of land.

In the early versions of his scheme, the site of the settlements was to be 'L5', a region on the orbit of the Moon where structures set travelling in the right orbit would stay put indefinitely, stabilised by the combined gravity of Earth and Moon and proof against disturbance by the Sun's gravity. 'L' stands for Lagrange, Joseph Lagrange being the eighteenth-century theorist who first realised that dual gravity creates such stable positions in space. As the concepts matured and other experts joined in, another kind of orbit came to be preferred, the so-called '2:1 resonant orbit', with the settlements swooping around the Earth every two weeks and the elliptical orbits swinging under the Moon's influence. In later studies still, any circular orbit more than half-way to the Moon seemed suitable. But while the precise methods and aims shifted, as O'Neill and his confederates continued their work, the central ideas and the essential enthusiasm remained undented. What follows is a sketch of O'Neill's scheme as it stood in 1977.

To justify the expense of establishing the first space settlements – the initial investment by the earth-dwellers – the original settlers will be employed in building satellite solar power stations. These will intercept sunlight over a huge area and supply enormous amounts of electrical energy to the Earth, thereby solving the world's energy problems. The power stations will be in geosynchronous orbit, circling the Earth every twenty-four hours and thus remaining poised over selected regions. They will transmit energy in the form of microwaves to large receiving areas on the ground, which will convert it into electricity. The power stations will be far enough out to remain clear of the Earth's

Inverted worlds in space, with rooftops pointing inwards, are conjured up in this NASA cutaway impression of a Bernal Sphere. Because the spinning of the space settlement produces artificial gravity acting outwards, the inhabitants live on the inside of the sphere.

They see houses on the far side of the settlement some 1,600 feet overhead. In O'Neill's scenario, the Bernal Sphere is a comparatively modest settlement for the early stages of manufacturing in space. (See also page 16).

Aboard Skylab *in 1973 an astronaut could take a shower – one of the first concessions to comfort in space. The space laboratories of the 1970s remained spartan, but ever-longer flights aboard them demonstrated the* *possibilities of living in space and carrying out important scientific experiments and observations – also of repairing spacecraft, like the* Skylab *itself, which was damaged in launching* (lower photograph).

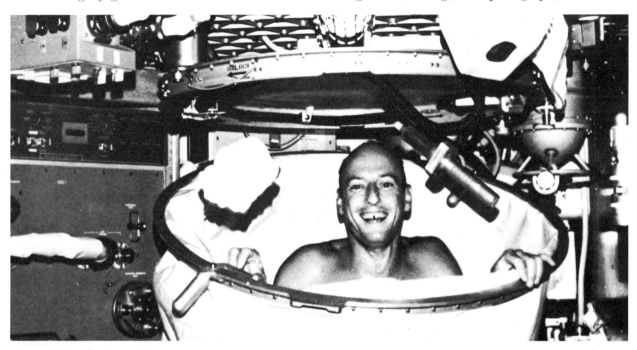

The Space Shuttle, *the recoverable and reusable space launching system under development by NASA, provides a practical basis for calculations about space colonisation. The upper photograph shows the orbiter* *flying piggyback off a Boeing 747 aircraft, for landing trials. In the lower cutaway painting, the orbiter is seen riding on the big booster that will launch it off the ground and send it on its way into space.*

'**Mass driver**' *is the name of the electromagnetic gun intended for shooting lunar materials into free space. In the working model demonstrated by the Massachusetts* *Institute of Technology a 'bucket' (upper photograph) travelled along a magnetic accelerator (lower photograph), faster than the eye could see.*

shadow and so will avoid the traditional problem of solar energy – what do you do at night and in bad weather? A satellite solar power station supplying 10,000 megawatts on the surface of the Earth will have a mass of about 80,000 tons.

Rather than attempting to build it on the ground and launching it into space, as some people had proposed, O'Neill wanted to make it from lunar rock, using a workforce housed in space. The relatively weak gravity of the Moon, and the absence of an atmosphere, meant that delivering materials from the Moon into Earth orbit would be much cheaper, pound for pound, than launching them from the Earth; it would also avoid harming the Earth's atmosphere with large numbers of big launching rockets. Thus, by 1977, building space settlements and building power stations went together in the plans of O'Neill and his colleagues.

In keeping with their policy of 'dealing in terms of the real world' they based their plans for getting equipment and people into space on an actual launch vehicle, the Space Shuttle, due to become operational about 1981. By the mid-1980s the Shuttle will be able to carry up into low Earth orbit as much as a couple of thousand tons of equipment a year. That would permit them to establish, within a period of about six or seven years of shuttle flights, systems for manufacturing in space from lunar materials, with an output of about 200,000 tons of finished product per year.

Another link with the 'real world' was a plan for a more thorough geological survey of the Moon, that would help to identify the best sites for mining. Local knowledge of the Moon's chemical composition came from the samples brought to Earth by American astronauts and Russian unmanned landers from particular spots on the lunar surface. They showed that important metals, together with bound oxygen, were abundant in the lunar regolith or 'soil'. Noticeably lacking were volatile materials, including water. There were differences in the composition of the rocks from different parts of the Moon – especially between those of the highlands and those of the flatter maria. Roughly speaking, the highland material was a better source of aluminium and glass and a poorer source of the iron and titanium that darkened the rocks of the maria. In the highlands there was a much greater thickness of fine-ground rock, easy to scoop up.

A start had also been made in making more widespread chemical maps from orbit in the *Apollo 15* and *16* missions. Under natural bombardment by atomic particles from the Sun, the surface of the Moon emits X-rays and gamma-rays, carrying the signature of partic-

ular elements, which can be detected by instruments on an orbiting spacecraft. The *Apollos* scanned about twenty per cent of the surface, mapping for instance iron and the radioactive elements. Space-scientists had a scheme for flying an unmanned mission, called the lunar polar orbiter or *Prospector*, that would be able to map the whole Moon with more advanced instruments. One hope for this mission was that it might find ice on the Moon. In most places the lunar soil appears to be baked and dried out by the intense sunshine of the lunar day. But near the poles of the Moon there are areas in permanent shadow, where the sunshine never strikes, leaving them cold enough for water to remain there for the lifetime of the Solar System. When *Prospector* peers inside those dark regions it may find both water and frozen carbon dioxide. Although the Princeton prospectus did not depend upon it, such a discovery would greatly enhance the possibilities for living off lunar materials, and probably help fix the site for mining operations.

The lunar material needed in the first phases of building space settlements and power stations can be scooped up by a single radio-controlled bulldozer, working the loose surface material. After a little sieving and sorting, the material will be packed into glass-fibre sacks made from moonglass, and then shot into space using a machine called a mass driver. As a way of moving material about, the mass driver, an electromagnetic gun, had an early and vital place in O'Neill's scheme. Henry Kolm at the Massachusetts Institute of Technology had been working on a system of high-speed ground transport using electromagnetism both for levitation and for propulsion. He brought to that development his long experience in using MIT's extremely strong magnets for research purposes. O'Neill persuaded him to try to develop a mass driver for space and Kolm was demonstrating a small prototype by 1977.

The device propels a special bucket, equipped with superconducting coils carrying a strong electric current. Each bucket rides like a surf board on the crest of a magnetic wave generated by circular coils arranged along a guideway. The bucket carries the sack, the rock or whatever material is to be accelerated; having reached the necessary velocity the bucket slows down, catapulting the payload along its trajectory. For retrieving material from the Moon, Kolm envisages a succession of buckets launching forty-pound loads at a rate of about one per second, with the buckets returning to the starting point by way of a loop.

The mass driver sitting at the lunar mine will shoot the sacks of lunar material towards L2, the Lagrange point closest to the Moon, which they will reach reliably even if there are small launching errors. Checked by the Moon's gravity, the sacks will arrive fairly slowly at a catcher stationed at L2. After collecting the lunar soil for a period of weeks or months, the catcher will travel to the manufacturing site in orbit about the Earth. The catcher will be propelled by another mass driver, but this time used as a reaction engine – it will hurl out some of the load to push the catcher along. The round trip will take several months, but half a dozen catchers will keep up a steady supply.

Arriving at the orbital space factory, the lunar soil will be turned into construction materials, and then into power stations and space settlements. Processes working at normal industrial temperatures will extract metals, including iron, aluminium and silicon, and will also release oxygen gas. Some of the iron, present in the lunar soil in the form of grains of metallic iron, is easily sorted by a magnet. The silicon will be a prime material for making the solar power station. Glass and ceramics can be made very simply by melting the lunar soil.

Apparently no special inventions will be required to build the necessary chemical processing and manufacturing plant that will operate in space. An early design for a prototype envisaged a plant of 300 tons mass that could accept and process about 1000 tons of lunar material a day, and turn out useful products equal to its own mass in less than six days. It would need 450 megawatts of power provided by solar energy. Scaling the plant up to five times that capacity will give a throughput of about 1,500,000 tons of material a year, and will require an estimated workforce of about 5000 people. Their spells of work away from the Earth might be six months or three years in the first instance; but in the long run people will spend their entire careers working in space.

O'Neill's idea of new communities in space gripped the imagination of many people apart from scientists and engineers. Stephen Cheston of Georgetown University, himself an historian, told us how, across the US, international relations experts, lawyers, sociologists and psychologists had taken up the concept. They established workshops and developed courses for their students. They were joined by people from the counter-culture, like Stewart Brand, publisher of *The Whole Earth Catalog*, and Timothy Leary, the prophet of ecstasy through drugs. And citizens' groups began to form, bringing together people from diverse sectors of American society, putting liberals, who ten years earlier had been trying to end the Vietnam War, alongside conservative industrialists.

Henry Kolm, engineer of the mass driver, made this comment:

'I know there are people, particularly many of my colleagues among physicists, who say this is no way to make people happy: they'll spend the rest of their days in a spinning aluminium cylinder. On the other hand I am surrounded by the most enthusiastic bunch of students I've ever worked with. There's no question that they'd volunteer to settle a space colony today if it were possible. It turns them on. I've tried to analyse the nature of their feelings. Of course there's a lot of élitism in it – they like to share their lives with an élite group. Some of them have said that the world is such a mess and the world is full of such scums that the chance to get away from it all is worth all the effort. One wonders whether that's what perhaps drove the Eskimos to settle the Arctic regions, or perhaps what made a farmer in 1630 settle in northern Maine.'

But the criticisms from other scientists and engineers were not only about human happiness. At the outset, O'Neill had contemplated enormous space settlements. The largest that he thought to be within the range of twentieth-century technology and materials would be fifteen miles in diameter, seventy-five miles long and with a land area of 7000 square miles – almost the size of Wales or New Jersey. He conceded, though, that such large habitats would be uneconomic and wasteful of materials. To some critics it seemed a flaw in the whole concept, that large settlements would be proportionately far more expensive to build than small settlements: for them, a sound engineering concept ought to bring the usual benefits of scale.

Detailed considerations of effort and cost steadily drove the thinking back to fairly small settlements, for a beginning at least. Yet a truly comfortable and self-sufficient space settlement ought to be as large as possible. The rate of rotation to produce a centrifugal force equal to the Earth's gravity will be correspondingly less, and with it the risk of motion sickness. And biologists raised doubts about whether small ecosystems could survive indefinitely – a possibly crucial point to be pursued in the next chapter. From the biologists' point of view, the problem will not be to build space settlements as large as O'Neill described, but to keep anything as small as that alive.

Objections that at first seemed minor began to accumulate as the studies proceeded, and each remedy added to the difficulty and costs of building a space settlement. For example O'Neill had thought at first that a rarefied atmosphere of pure oxygen would sustain life on the colony. Later schemes allowed for some nitro-

The loose lunar soil *or 'regolith', rich in useful materials, can be simply scooped up. A single bulldozer operating at a lunar mine (model,* upper *photograph) could meet all the initial needs of space manufacturing. Another* model (lower photograph) *shows a lunar mine and the start of the long mass driver intended for shooting sacks of lunar soil into space, at the start of the sequence continued on the next two pages.*

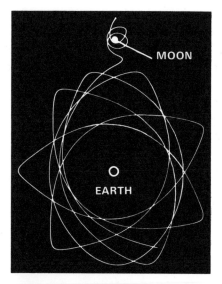

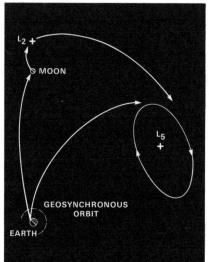

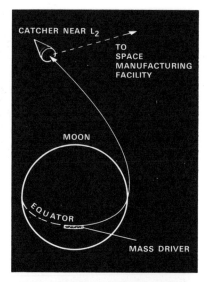

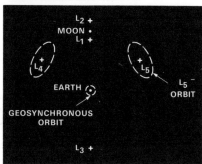

Early ideas about orbits *for space settlements and manufacturing sites appear in two diagrams* (top left and centre). *The first shows materials en route from a collection site near the Moon ('L2') to a highly eccentric orbit ('2:1' resonant orbit) carrying the settlements around the Earth twice a month. In the second diagram, materials flow instead to a region in the wake of the Moon ('L5') where the settlements would circle the Earth once a month. 'L2' and 'L5' are among the Lagrange points* (lower left), *regions of gravitational stability in the Earth–Moon system. Later ideas about eventual orbits are less exclusive, but L2* (diagram, top right) *still provides a valuable region for collecting sacks of raw material, to be shot from the Moon and trapped in a 'mass catcher'* (model, right).

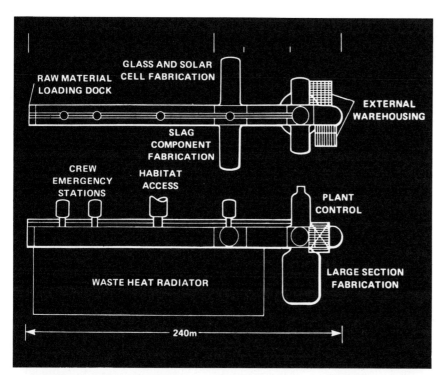

RAW MATERIAL LOADING DOCK

GLASS AND SOLAR CELL FABRICATION

EXTERNAL WAREHOUSING

SLAG COMPONENT FABRICATION

CREW EMERGENCY STATIONS

HABITAT ACCESS

PLANT CONTROL

WASTE HEAT RADIATOR

LARGE SECTION FABRICATION

240m

A factory in space would take in lunar materials at one end (delivered by the mobile 'mass catcher') and deliver products or parts at the other end. Methods of chemical processing and fabrication were worked out in some detail in collaborative studies by O'Neill's associates and NASA. A space manufacturing facility of this kind might be directly attached to a comfortable space settlement, or it might orbit independently like the model (below). It could process material equal to its own mass, every few days.

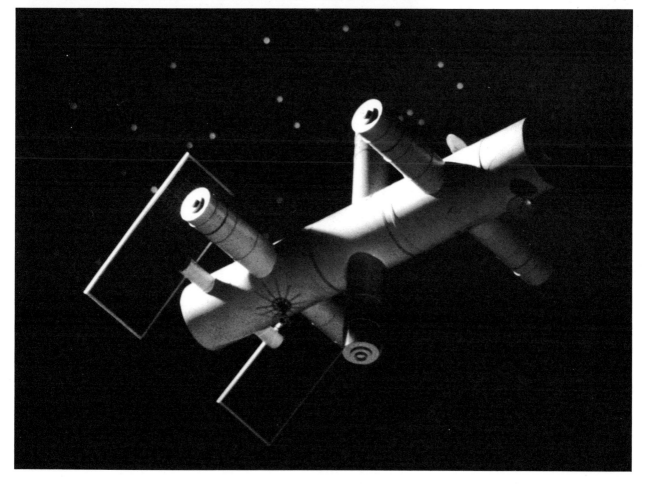

gen, though not as much as on Earth. The pressure on the container increased accordingly, bringing severe penalties in mass and cost. It also required the transport of nitrogen from the Earth's surface. One reason for adding nitrogen was to avoid peculiar effects on the human respiratory system; another was to retain something like the normal interchange of nitrogen between the atmosphere and soil bacteria. Furthermore, even quite a small settlement will need millions of tons of shielding against cosmic rays, to be added around the basic structure – a requirement which O'Neill underestimated at first. Fancy ideas about building a space settlement in the form of an elegant torus or doughnut evaporated when the problems of shielding were taken into account.

With such snags besetting him O'Neill was forced by 1976 to offer a cut-price design for a space settlement, called the Crystal Palace. Instead of a big hollow container filled with an atmosphere, as in the earlier versions, it will consist of two connected structures, each very like a stack of bicycle wheels. That is to say, the atmosphere will be confined to tubes running around the circumference, while spokes hold them in shape. One 'stack' will make up the living accommodation; it will have Earth-like pseudo-gravity and be fully shielded against cosmic rays. The agricultural areas in the second 'stack' will be scantily shielded and the gravity will be less.

With headroom of, say, a hundred feet, the living areas will lack the aesthetic appeal of the Bernal Sphere. But the Crystal Palace seems to be the cheapest way of providing a flat, land-like surface in space. It will be luxurious compared with the shacks proposed for the first couple of years' work on satellite solar power stations: for them, tanks shed by the Space Shuttle will simply be fitted with floors and the essentials to support life in zero gravity.

Despite all the difficulties O'Neill remained optimistic about his overall scheme, and its costs. He told us:

'We've estimated – this has now been done by a number of different studies – that for something like the cost of the Apollo Project we could set up a large-scale manufacturing facility in space which would already be able to supply a large fraction of all of one country's needs for new electrical capacity. And if the cost of the Apollo Project sounds like a lot, then I should tell you that the amount that the electric utilities in the United States alone are going to have to spend on some kind of new generator capacity over the next twenty-five years is about equivalent to forty or fifty Apollo projects. So we're only asking for a few per cent of that.'

And it could all be happening by the 1990s. But not everyone believed so. Val Cleaver, a British rocket engineer, said shortly before his death in 1977 that he was flabbergasted that a distinguished physicist 'who ought to know better' was seriously suggesting that we could have space settlements on the time scale and for the amounts of money that O'Neill was talking about. Cleaver was full of admiration for a study that might become a classic of astronautics, but he suggested that the difference between O'Neill's most modest settlement and the space stations of the 1970s was like the difference between a dinosaur and a mouse. From *Sputnik* to *Apollo* was a very small step compared with the further step to a vast floating community. Engineering took time – that was Cleaver's main point. A leading US senator, William Proxmire, was even blunter. 'Not a penny for this nutty fantasy!' he declared.

The weakest point of O'Neill's scheme will, I suspect, turn out to be his adoption of the satellite solar power stations as the primary justification for the settlement of space. Energy became the fashionable concern of the 1970s, after the Arabs priced their oil more realistically to take account of its impending exhaustion. But that very concern prompted a flood of technical and economic studies of future energy supplies. They mapped out the wide range of possible energy sources – conventional, nuclear and unconventional – with which satellite solar power stations would have to compete. In June 1978, despite Proxmire, the US Congress augmented NASA's budget for research on satellite power, and asked another agency to consider whether the stations might be built from lunar or asteroidal materials. The votes did not guarantee a favourable outcome.

Recall all the activity of mining the Moon, creating habitats in orbit (however spartan in the first instance) and building the satellite solar power stations. The upshot would be to deliver to the microwave 'farms' on the ground radiant energy half as intense as sunlight at the same places. Indeed this figure was given as an argument for lack of harm to the environment! Even allowing for the day and night reliability of solar energy in space, and for a highly efficient conversion of microwave energy into electricity, would it not be a better bet, the critics said, to concentrate on improving the direct conversion of solar energy at the Earth's surface?

His preoccupation with putting many people into space as quickly as possible may have caused O'Neill both to overstate and to understate the possibilities before us. In particular the concept of large workforces in space, necessary to sell the idea of space settlements, lured him into a possible blind alley with the solar

'Crystal Palace' *is a cut-price design for providing reasonably Earth-like conditions during the earliest phases of space colonisation. Unlike the Bernal Sphere (see illustrations pages 16 and 31) in which the whole interior is filled with an atmosphere, the Crystal* *Palace confines its atmosphere in tyre-like tubes containing the surfaces where people and crops actually live. To accommodate 6000 people, Crystal Palace would have a basic mass of 49,000 tons, but would need three million tons of shielding. QE2 to scale below.*

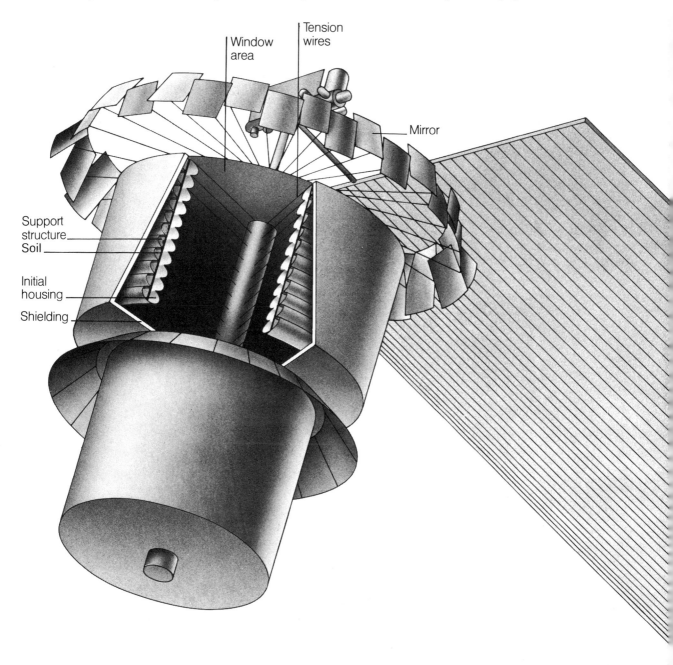

satellite power stations, as he tried to think what the workforces might actually do, apart from building more settlements. The thoughts of other experts, notably Theodore Taylor, ran in the direction of essentially unmanned space factories, leaving it to robot machines as far as possible to cope with the unusual environment in space – and eventually to build comfortable space settlements for people.

Even among experts who favoured ever-increasing human activity at the 'high frontier', to carry various space ventures forward in permanent bases on the Moon and in orbit, we met several who urged on us an alternative scenario. Living conditions, in their view, will remain for a long time more like the *Skylab* of the 1970s and the *Spacelab* of the 1980s than *Bernal Alpha* of O'Neill's imagining. Then the time will come when space observatories, space mines, space factories, space transport systems and (one must add) space navies will be sufficiently populous for the people to start getting together to build roomier cities in space.

Along the way many lessons will have been learned about how to build and how to survive in space. For example it will take years, by definition, to test the durability of materials in long exposure to the combined effect of vacuum, low gravity, ultraviolet light, solar particles and the thermal shock that a spacecraft feels every time it turns. The tough frontier experiences on the Earth are a precedent for the step-by-step approach, rather than for O'Neill's pictures of comfort and safety to be achieved at quite an early stage by urgent methods. As for the idea that action in space will soon be solving the world's problems, David Brower of Friends of the Earth spoke for many when he told us: 'If something's going wrong with this planet we'd better fix it here and not look for some sort of escape.'

However it unfolds in practice, O'Neill's vision will be fulfilled, I think, as people exploit unused resources of the Solar System. As wealth in space grows, the space cities will come, with all their capacity for exponential multiplication. In time the population in space will indeed outstrip the population on Earth – eventually far outstripping it, as the Dyson Sphere begins to become visible. And eventually, as O'Neill has predicted, some of our descendants will spend their lives travelling in spacious settlements from star to star.

O'Neill's big idea, first and last, is a scheme for transferring life from Earth to space and ultimately through huge deserts of the universe. For some critics, not at all opposed to that general objective, O'Neill and his engineering colleagues have seemed altogether too ready to adopt the agriculturalist's stark view of life. The space settlements of the future may have to be more like Noah's Ark than a model farm – and therefore more grandiose.

4

A TALE OF
TWO PLANETS

The physicists were talking about taking *life* into novel environments in space. But the people who understood life and what its special qualities and needs might be were the biologists. Even while Gerard O'Neill was conceiving his space settlements and cheerfully envisaging pest-free crop production in space, and astronomers were trying to estimate the number of inhabited planets in the Galaxy, a new hypothesis about the character of living systems was developing. It promised to give a clearer idea of what might live, what might die, and what might never come to life unaided. Here I

relate how the comparison of two planets, Mars and Earth, prompted the hypothesis, and tell of some of the possible implications for living in space and living on Earth.

The two *Vikings* landing on Mars in 1976 carried what the optimists called 'biology instruments'. Nobody expected any encounter with the martian engineers who allegedly built a network of huge canals. The close-up pictures from *Mariner 9* orbiting Mars in 1971–72 confirmed once and for all that the 'canals of Mars' were a visual illusion afflicting otherwise competent astronomers on Earth. There were natural formations resembling dried-up rivers, but the pictures showed dust-blown landscapes devoid not only of canals but of any territory promising for life. Mars looks like a rocky desert, and three Russian landers wrecked themselves among the rocks.

Yet for the *Viking* experimenters microbes in the martian soil would have been ample reward – the merest symptom of life that would tell us we weren't quite alone in the Solar System. The discovery of forms of life subtly different from the Earth's would have allowed biologists to explore afresh the fundamentals of life. Especially they hoped for confirmation that life would tend to originate wherever the conditions were not impossibly severe.

The pictures from the martian surface confirmed the earlier, more distant, impressions of a rocky desert. Nothing moved. As the automatic laboratories radioed their reports to Earth, some short-lived mysteries in the data produced by chemical reactions in the martian soil kept the hopes of life-detection flickering for a few weeks. Eventually it was clear, however, that no martian microbes supped the food or absorbed the carbon dioxide offered to them. As the equipment would certainly have detected life in any terrestrial desert, it required dogged optimism to suppose that similar searches in other places would find life. The possibility remained that Mars was once alive, if and when rivers ran, but the provisional verdict from the *Vikings* was that Mars was stone dead.

Not everyone was disappointed. People looking forward to the colonisation of Mars by human beings could be relieved that there was no sign of aboriginal inhabitants; the presence of the simplest life would have made Mars a nature reserve and quarantined area for centuries to come. Nor did the outcome displease those scientists who suspected that life was an altogether less probable phenomenon in the universe than the conventional wisdom of the times would have it. James Lovelock, a British scientist doing work for the US National Aeronautics and Space Administration, had special grounds for satisfaction. He irritated his colleagues by promising them that Mars was dead, long before the *Vikings* set off.

Lovelock was a rare bird in the late twentieth century: a successful freelance scientist, operating from his home in a Wiltshire village. He based his business and his research upon his invention in the late 1950s of one of the most sensitive chemical detectors ever made. In it, molecules reveal their presence by capturing electrons released by an electric discharge. You can then take gases or vapours, separate the constituents by 'gas chromatography' according to their speed of travel through a column of adsorbent material, and use Lovelock's electron-capture detector to identify the constituents: they show up even if present only in extremely small amounts, as environmental pollutants for example. One of the early applications was in detecting the traces of DDT that were scattered through the Earth's environment.

In the 1960s NASA invited Lovelock, as a hunter of trace materials, to help in the design of experiments for detecting life on Mars. It occurred to Lovelock that if life was present it ought to affect the composition of the martian atmosphere. When he looked back, as it were, from Mars to the Earth, he was struck by what a strange place the Earth is. The Earth's atmosphere possesses unstable mixtures of gases, including for instance the highly combustible hydrogen and methane in the presence of oxygen. Mars, on the other hand, seemed to have an altogether more predictable atmosphere; just the sort you would expect if nothing but the laws of physics and inorganic chemistry held sway and any chemical instabilities had been eliminated. It was the unsurprising nature of the gases of the atmosphere of Mars that provoked Lovelock's pronouncement that the planet would be found to be dead.

That was not the end of his train of thought. When Lovelock looked more intently at the Earth and its atmosphere he began to doubt another conventional opinion of the time. It was the notion that life on Earth had been lucky, that living organisms on Earth were able to evolve steadily over thousands of millions of years because the conditions of temperature and chemistry remained always congenial to life. Was it just luck, he wondered, that small changes in any of a dozen critical factors which could have killed everything did not occur in all that long history?

A more sophisticated version of the good-luck theory credited elaborate physical and chemical regulators at work among the loose stuff at the planet's surface: for example, the saltiness of the oceans eventually depended upon the great processes of continental drift and mountain building. Occasionally great crises of life did occur, like those that wiped out the trilobites 225 million years ago and the dinosaurs 65 million years ago. But usually conditions remained stable for life. To Lovelock, the chemical regulators seemed insufficient to account for the near-constancy of the temperature and atmospheric composition of the Earth.

He stood the issue on its head, with the supposition that life itself controlled the temperature and chemical

The Gaia hypothesis, *that organisms of all kinds collaborate unwittingly in keeping the planet Earth fit for life, originated with a chemical analyst, James Lovelock (left). A cell biologist, Lynn Margulis (right), helped in* *portraying the special role of micro-organisms. The most important sites of gaian regulators seem to be shallow waters – for example, estuaries like the Blackwater estuary in England (bottom).*

environment to ensure its own survival, even in the face of great changes in the intensity of energy coming from the Sun. It was not luck, Lovelock suggested, but unconscious cooperation among living organisms, from microbes to elephants, that kept the Earth in the peculiar state in which all species could prosper. Thus living matter, together with the air, the oceans and the continents, made up a giant and complex system which behaved almost like a living organism that maintained its internal milieu by subtle feedback processes. To this hypothetical creature Lovelock gave the name Gaia at the suggestion of a Wiltshire neighbour, the novelist William Golding. Gaia (pronounced gayeeah) was the Earth-goddess of the early Greeks.

Great feedback processes in the biosphere were already well known to scientists: for instance, the way plants took in carbon dioxide and gave out oxygen as they grew, while animals took in oxygen and gave out carbon dioxide; or how some microbes fertilised the soil by 'fixing' nitrogen from the air, while other bacteria returned the nitrogen to the air. Dozens of such cycles were well recognised. But the Gaia hypothesis assigned a much more positive and systematic role to organisms and to their unconscious knack of keeping things just so. Gaia might always remain more of a metaphor than a scientific hypothesis open to formal tests, yet it was a fruitful source of new interpretations and experiments.

The idea became sharper when an American cell biologist, Lynn Margulis of Boston University, joined Lovelock in the early 1970s as a votary of Gaia. Margulis already had a reputation for unconventional thinking. When I first met her she was preoccupied with the evolution of life during the long age of the microbes, when only the simplest organisms lived on Earth and the chemical relationships between life and the planet were just establishing themselves. She was a forceful advocate of the theory that the 'modern' living cells of plants and animals evolved by the clubbing together of cells of primitive microbes. For example, when the growth of blue-green algae first let free oxygen loose on Earth, it was a deadly poison. In the resulting 'oxygen crisis' certain bacteria evolved which were capable of dealing with the oxygen. As Margulis argued, it was advantageous for other microbes to take these bacteria in as permanent lodgers. The descendants of those bacteria are still to be found living symbiotically in the cells of the human body, where they are called mitochondria and continue to cope with the oxygen. They are living evidence of a long-lasting collaboration between organisms to preserve the conditions of life.

To Margulis, Lovelock brought the problem of what could be maintaining the 'anomalous' gases in Earth's atmosphere in the presence of oxygen. She was able to offer a long list of micro-organisms that produced hydrogen, methane, ammonia and so on. They did so at a great cost in energy, ultimately derived from the energy of sunlight. In other words, Gaia was working hard to maintain the peculiarities of the Earth's air. In Lovelock's theory, the air acts like the bloodstream of an animal, bringing essential supplies to all of the Earth's inhabitants and carrying waste products away. Countless numbers of humble and unregarded micro-organisms play their parts in keeping this 'bloodstream' purified and replenished.

Gaia's key controls seem to be located in relatively neglected parts of the Earth's surface – in marshlands, in the mud at the bottom of estuaries and in the waters of the continental shelves. Dry land and the wide open oceans are not as important as their large areas might suggest. For example, the airless conditions in relatively small bogs and enclosed seas have deposited and buried carbon fuels, thereby helping oxygen to persist in the atmosphere of the whole world. Yet the proportion of oxygen in the Earth's atmosphere does not climb above twenty-one per cent. A student of Lovelock's found out that, should it rise by only a few per cent, vegetation would become so inflammable that even the tropical forests would burn to ashes. That might be a drastic way of consuming oxygen and restoring the status quo, but Gaia seems to prefer to keep the potentially dangerous oxygen within bounds. Just how the trick is done remains to be discovered.

As for suggestions that peculiar microbes will be found somehow contriving to live in peculiar circumstances in the atmospheres of Venus or Jupiter, or in some forgotten corner of Mars, Lovelock remarked in conversation with us: 'There is no *sparse* life – a planet is either living or not, just as a person is either living or a corpse.' Margulis too was sceptical about finding life on other planets. She described life as a gentle phenomenon, requiring mild conditions and constant exchanges of gases with its environment, under moderate pressures and temperatures that did not seem to exist on any other planet. 'If there is going to be life elsewhere in the Solar System,' she told us, 'it is really going to be us.'

The gaian picture of the Earth and its intricate biochemical control systems will veto simple-minded engineering of settlements in space. Anyone tempted to regard the air as merely a mixture of certain gases, which can be taken along in bottles, will come to a sticky end. According to Margulis we shall have to carry substantial pieces of Gaia along with us, if we wish to

IG119 PDP-1 SDL-30 CLR CUSTOM 10P240 L1200A GRE-2 9-1-76
IPL PIC ID 76/08/31/195505   WIDB/L1200AX
JPL IMAGE PROCESSING LABORATORY

No martians *admire the sunsets on the Red-Planet. The picture (top) was transmitted from Mars to the Earth by the* Viking 1 *lander. The unearthly scene resembled, if anything, a dusty sunset over a barren desert. The* Viking *landing robot is depicted (right) in a terrestrial photograph. Two such landers parachuted safely to the surface of Mars in the summer of 1976. A picture transmitted from Mars (above) shows the sampling arm in operation on the martian surface. The landers scooped up samples of the soil and tested them for evidence of microbial life in small, automated laboratories inside the vehicles. Excitement flickered for a few weeks among the scientists in the Viking biology team, because some of the tests gave positive results. But chemical activation of the martian soil by ultraviolet*

light from the Sun was a plausible explanation of the positive results. By 1977
the consensus was that the martian soil
was barren – at least in Chryse Planitia
and Utopia Planitia, the regions visited
by the landers. A morning picture from
Viking 1 (above) gives an exceptionally
vivid impression of a rock-strewn desert
on Mars. The foreground object in the
centre is the boom supporting the
Viking's weather station. The large,
sharp-edged features to the left are
sand-dunes, and piles of sand to the
right of the boulders indicate a recent wind-storm. The Jet Propulsion
Laboratory, California, which master-
minded the US exploration of Mars,
prepared the general view of the planet
(right). The laboratory used a computer
to manipulate more than 1500 individual pictures sent from Mariner 9
orbiting Mars in 1971–72. Conspicuous
features include the 80,000-foot volcano
Nix Olympica, lower centre, and the
concentric rings of the north polar ice-
cap at the top of the picture.

survive comfortably in space for long periods. Consider the simple experiment of taking water, aquatic plants and fishes from a pond and putting them in a box. Within a week there will be a stench and the fishes and plants will be dead, with only a microbial scum surviving. Similarly, small isolated systems of life in space will be unstable and will die easily.

In the 1970s nobody knew just how large a plot of land, air and water was needed, for indefinite survival. Large systems like forests and lakes were evidently stable for hundreds or thousands of years. Scaling up systems to the size of an ocean provided stability for much longer periods, but scaling down seemed fraught with uncertainty. According to Margulis, it will be a 'terrific challenge to the mind of man' to set up a system that will be liveable in hostile places. Indeed one of the great benefits of the space enterprise will be in forcing attention to the stability of small ecological systems, and requiring many experiments to be run on Earth before they are hazarded in space. The human love for green and pleasant landscapes may be thoroughly vindicated as a guide to what is necessary for survival. In the upshot, as Margulis suggested to us, space settlers may do well to imitate the gardener on Earth. He does not know the names of all the bacteria and fungi at work around the leaves and roots of his plants, but maintains a familiar and pleasing system, provides water when necessary and generally works to 'keep it alive'.

Another microbiologist, John Postgate of the University of Sussex, echoed Margulis's advice about the need for large systems to support life in space. Postgate became a leading authority on the genetics of nitrogen-fixing bacteria, after working in quite different areas of microbiology. I knew him to be interested in all aspects of the life of microbes, and their role in human life. He told us that any notion of a sterilised, man-managed system in a space settlement will be as unattainable as it would be undesirable. Human beings carry a menagerie of 'germs' in their mouths, intestines and skin, and these micro-organisms are important for their health. People and animals kept in a confined environment for a long time will tend to become colonised by only one strain of micro-organism and that may be damaging – for example, in the loss of micro-organisms making certain vitamins needed for life. It will be essential to take into space pretty well all the micro-organisms with which we are normally associated on Earth. Although bacteria are very small, they will also need the environments to which they are adapted. Only by starting with a large space station will you be able to take sufficient soil, marsh, plants and animals to

set up the right sort of ecological system to keep going.

The release of intestinal 'wind' might be fatal for the inhabitants of an ill-conceived space settlement. The micro-organisms in human intestines produce methane; and in experiments with astronauts living in an enclosed space the concentration of methane has reached one per cent in just a few weeks. That concentration is safe enough but Postgate warned that the methane could eventually become explosive. An obvious solution will be to include in the space settlement representatives of the bacteria which on Earth convert methane into carbon dioxide.

The *Bios 3* experiments by the Soviet Academy of Sciences, I should add, confirmed the difficulties of running very small, closed systems. In a simulation on the ground of a manned space environment, reported in 1975, three men spent 180 days in *Bios 3*, sharing it with algae (*Chlorella*), cereals and vegetables. Some results were good: the atmosphere seemed self-sustaining as far as human needs were concerned, nearly all the water was successfully recycled, and the 'bionauts' obtained almost a third of their food from their sealed garden. They could not digest the *Chlorella* as intended, and there were subtle changes in the trace metals present in the crops. In a related experiment, an uncontrolled mass of algae surrounding the roots of the cereals and vegetables slowed their growth. But one enigmatic part of the report from the Russian scientists bore out all too strikingly the misgivings about small ecosystems. In the 180-day experiment, the air shared by the crew and the *Chlorella* became toxic to the vegetables. Potatoes and tomatoes stopped growing and the leaves of cucumbers were discoloured. The poisonous agent in the air was not identified but, when the Earth's air was reintroduced, the plants in question quickly recovered.

Turning to the prospects for the Earth itself (or 'herself' as Lovelock prefers to say) what can we say about the relationship of human beings to the regulatory cycles of materials through the Earth's air? Both Lovelock and Margulis told us that they thought that the control systems were generally robust enough to cope with human activities. As Margulis pointed out, the 'oxygen crisis' on the early Earth was a much more severe challenge than any man-made pollution. Gaia could cope with us and survive – though we might be extinguished in the process. Lovelock took a sceptical view about the 'environmental revolution' that his sensitive instruments had helped to spark. He himself had detected the halocarbon propellants of aerosol sprays in the air of western Ireland, which started another furore. But, pursuing his investigations, Lovelock found that nature was making similar compounds and destroying them on a far vaster scale than the human chemical industry. One of the natural agents which Lovelock found busily transferring iodine back from seaweed to the land was methyl iodide – widely condemned as a man-made cancer-causing chemical.

Not that Lovelock favoured complacency about Gaia. Our species should take care not to damage the natural control mechanisms or reduce their ability to cope with unforeseeable natural calamities. If we are wise, we shall postpone any farming of the mud of the shallow coastal waters until we know that repository of gaian safety much better. And one particular pollutant may yet be capable, in Lovelock's judgment, of overwhelming the natural regulators: carbon dioxide. During the nineteenth and twentieth centuries, the burning of coal and oil and the destruction of forests have caused a rapid increase in carbon dioxide in the atmosphere.

In a hundred years the carbon dioxide in the air has increased by ten per cent. Let me comment on that. Standard atmospheric theory says that carbon dioxide in the air warms the Earth. It acts like a greenhouse window, letting in sunlight while interrupting the radiation of heat from the ground back into space. Despite a nagging doubt that the effect might go the other way, by increasing the cloud and so cooling the Earth, meteorologists in the 1970s generally thought overheating to be a danger for the twenty-first century, while our species uses up the last accessible reserves of coal and oil. Nature's planetary control systems have been doing their best to absorb much of the excess carbon dioxide into the oceans and perhaps to accelerate the growth of plants, but it is evidently not enough. A moderate increase in the global temperature might well make the sub-tropical deserts and Europe wetter, while reducing rainfall on the great plains of North America. A severe increase could, in theory at least, cause a fatal runaway process in which the heated oceans shed their dissolved carbon dioxide into the atmosphere.

Freeman Dyson suggested to us that mankind should be ready with plans to check the increase in carbon dioxide by growing a great many new trees all over the world, in new places, to mop up the excess carbon dioxide. By his calculation, if every man, woman and child on the planet plants a tree before breakfast every day for a year, that will just about counteract the present rate of increase of carbon dioxide in the atmosphere. It would be costly but, in Dyson's opinion, a valuable exercise in international cooperation. As long as the trees are growing in the ground – say for a hundred years – they will lock up carbon and produce a

bank of energy which can be drawn on later by burning them. Any other plants will do, provided they are not allowed immediately to decay. The water hyacinth growing in tropical swamps and rivers will remove carbon dioxide from the atmosphere about eight times faster than a corresponding area of trees; sugar cane will be another fast-growing candidate. People may want to create artificial peat bogs where the harvests of such plants will accumulate and not be allowed to return carbon dioxide into the atmosphere.

The energy released by industry and power systems may also affect the climate. Experts foresee that the use of energy on Earth may be limited in due course when that 'anthropogenic heat' becomes significant compared with the input of energy from the Sun. Man-made heating is already perceptible locally in cities and industrial regions, but globally it seems insignificant – the natural energy flow from the Earth into space is about five thousand times greater. But if the population grows and world-wide living standards rise to the present American level or beyond, the effects may become serious, especially when compounded with the increasing carbon dioxide. According to Gerard O'Neill, one of the basic reasons for wanting to start shifting human life and industry into space, and for supplying the Earth with electricity from satellite solar power stations, will be to bypass that limit.

Looming behind these anxieties about overheating the Earth is the contrary and in some ways more definite danger that, for the next 60,000 years or more, our planet will be in the grip of another ice age. The gaian control mechanisms do not prevent such events. If nature takes its course, Alaska, Canada and large parts of northern Europe and Russia will be obliterated under sheets of ice a mile or so in thickness, as has already happened about twenty times in the past two million years. The effects will be world-wide. How soon and how quickly the next ice age will begin to bite in earnest no one can say, but it may occur fairly quickly, with the snows of winter failing to melt in the summer over wide areas in the northern lands. It will be a drastic change in the familiar water cycle of the Earth, enough to lower the level of the oceans markedly.

Predictions of this kind became possible in the 1970s when geologists, meteorologists and climatologists agreed at last upon the cause of the comings and goings of the ice. It became clear that the warm intervals between ice ages were brief – about 10,000 years. And from among dozens of rival theories one emerged triumphant, when ocean scientists produced a much more precise record of the past ice ages from the evidence contained in fossils of small animals in the mud of the ocean floor. The ice-age cycles fitted very closely with changes in the summer sunshine of the northern hemisphere, brought about by subtle but predictable variations in the Earth's orbit around the Sun, and in the tilt of the Earth's axis. A favourable astronomical situation lifted the Earth out of the most recent ice age 10,000 years ago, but since about 5000 years ago the trend has again been downwards. The next ice age will be the biggest natural challenge to be faced by our species since the time, 18,000 years ago, when our hunting ancestors were enduring the worst of the last ice age.

If we are in poor technological shape to meet it – if, for example, there is a general reversion to a peasant existence – the suffering will be terrible. But, given ways of keeping ourselves warm and growing food in difficult circumstances, general survival will certainly be possible in the ice age. And if there is confidence and skill in technology, it may well be possible to keep the ice at bay. The concern about overheating, as a result of the increase in carbon dioxide and the industrial release of energy, indicates that our powers are not necessarily insignificant. The resources of accessible coal and oil are far too small to maintain a special carbon-dioxide blanket indefinitely, but if we can secure unlimited supplies of energy, either from nuclear fusion or from space, we shall in principle be able to use them to melt the ice. For example, space engineer Krafft Ehricke of North American Rockwell, suggested in 1977 that huge mirrors in orbit about the Earth could serve to avert an ice age by directing additional sunshine on to the crucial zones where the ice tends to accumulate. But the effects on meteorological processes and wild life of sunlight coming from unusual directions would require careful assessment.

Whatever measures are adopted, it seems to me that holding back the ice will call for a lot of nerve. It is not just a matter of walking a tightrope between too much heat and too much cold. If you aim simply to maintain a climate like that of the twentieth century, you must expect some melting of the existing ice sheets, which are not yet in equilibrium after the last ice age. Many coastal plains and cities will be flooded. To win and sustain agreement on the procedures from the whole species will demand unprecedented good will. The prevalent objections to all schemes of climate modification are twofold: their consequences are very hard to predict with any confidence, and international uproar about actual or suspected harmful effects is far more probable than the sober agreements that would be necessary for

concerted action. In the 1970s some meteorologists were wondering if the workings of the world's weather and climate were not only complicated and paradoxical, but perhaps impossible to analyse completely because the atmosphere was fundamentally unstable.

One distinguished meteorologist sharing that suspicion was Jule Charney of the Massachusetts Institute of Technology. He was one of the chief initiators of the current efforts by meteorologists to combine computer models of weather and climate with global experiments for detailed data-gathering, to understand the natural processes as well as possible. But, as he remarked to us, ignorance about the smaller scales of atmospheric motion infects the very large scales. Charney also admitted to a prejudice against radical attempts to alter weather and climate, based on his perception of human life. Unpredictable weather and climatic changes had always presented interesting challenges and human beings might not have evolved without them. He was disturbed by the notion of space settlements with fully controlled environments, because human beings needed the vagaries of weather. And rather than try to prevent the next ice age, Charney thought that people should learn to live with the ice sheets.

On the other hand, where human activity has impaired the climate, Charney was willing to consider remedies, in particular to halt and reverse the increase in the area of deserts. In the 1970s a serious drought was affecting the Sahel along that south Saharan edge, and Charney reasoned that destruction of vegetation 'whitened' the land and helped the desert to advance.

Careful cultivation of the land at the southern edge of the Sahara, making the ground darker again, will perhaps halt and reverse the desertification in the Sahel. Similarly, where the northern border of the Sahara reaches the Mediterranean Sea between Tunisia and Israel, it may be possible to cure the paradox whereby winter storms penetrate into the desert yet drop very little rain. According to Charney's speculation, planting a strip of greenery a hundred miles wide along the northern borders of the desert will fool the air into thinking that it is still over the Mediterranean, so that it will continue to release its rain.

A quite different kind of development with planet-wide implications may come from an effort of biologists to use genetic engineering to make the world more fertile. Their aim is to enhance the biological processes that fix nitrogen from the air. This brings us back to the microbes of Gaia and the great cycles of atmospheric gases involved in life. In an extraordinary gesture of mutual dependence among species, evolution left the tasks of providing nitrogen nutrients for all the world's plants and animals to a few species of bacteria. Bathed in an atmosphere consisting mainly of gaseous nitrogen, we are quite unable to make any direct use of it in that chemical form. Only the nitrogen-fixing bacteria, living in the soil or in the roots of certain plants like beans and clover, are able to absorb the nitrogen gas from the air and turn it into a form that plants can assimilate. They fertilise the soil, and their output of fixed nitrogen sets a limit to natural biological productivity.

The introduction in 1914 of an industrial method of extracting nitrogen from the air, to make ammonia, brought about a revolution for those farmers who could afford the man-made fertiliser. By the late twentieth century the fertiliser factories were fixing 30 million tons of nitrogen, perhaps as much as a quarter or a half of the rate of natural fixation for the whole planet. But the industrial process worked at a high temperature and took a good deal of energy. The cost of ammonia fertiliser more than trebled in the 1970s, hitting hardest the farmers in the poor countries whose needs were greatest. Mere bacteria could do the trick at ordinary temperatures, drawing directly or indirectly on solar energy. Chemists found catalysts that would also work at ordinary temperatures.

John Postgate, whom I introduced earlier, set out to use genetic engineering to endow more species with the capacity to fix their own nitrogen in the biological fashion. In his experiments in the 1970s, at the University of Sussex, he and his colleagues succeeded in transferring the genes for nitrogen fixation from one species of bacteria to another. Thus they created new species of nitrogen-fixing bacteria. The next objective was to try to make crop plants that could fix nitrogen for themselves, without relying on an uncertain relationship with soil bacteria. As Postgate commented:

'Since we have this kind of genetic package with nitrogen-fixation genes on it, we're at least in a position to start trying to put these genes into plants. We know perfectly well what will happen the first time that we do this. The plants will not fix nitrogen because they'll want other genetic information. But as the years go on – and that's what research is all about – we hope to find out what the extra information will be.'

Postgate did not rule out the possibility of failure – that there might be some good reason why the nitrogen-fixing plant could not be attained. But success in creating, say, a nitrogen-fixing wheat may transform the prospects for the world's poor, struggling to feed themselves. In my opinion it may also mark a turning-point in mankind's management of the planet.

If people are to transplant life into the universe, and survive indefinitely out of reach of the Earth, they will need all the biological sophistication they can muster. As the Gaia hypothesis emphasises, life on Earth is wrapped up in complex processes involving the inanimate rocks, water and air, and also the climate. A very long evolutionary history has ensured that those complex processes work and space engineers cannot expect easily to match them. But interplanetary research itself may hold the key. By the 1970s, the very different kinds of weather and climate observed on Venus, Earth, Mars and Jupiter were already helping meteorological theorists in their efforts to understand the Earth's climate and to establish general principles of climate beyond the Earth. Although there is unlikely to be any other biological system for comparison, Lovelock and Margulis look forward to detailed studies of Mars and Venus. The results may tell us, for example, whether the Earth started very much like those other planets and then went its own way, or whether there was some crucial difference in the Earth all along.

It seems to me that the *Vikings* on Mars may have accomplished for planetary science something analogous to HMS *Beagle*'s contribution for natural history. Through the eyes of the young Charles Darwin, the voyage of the *Beagle* found living things adapted in Bible-defying detail to the niches available on Earth. The *Vikings*, with television cameras and chemical instruments, sampled a planet utterly different from the Earth and put a large question-mark against easy assumptions about the nature of life. Whether or not the Gaia hypothesis itself survives, the contrast between Mars and Earth will remain for all time one of the great object lessons of nature. As Margulis summed it up: 'If we were just an ordinary dead planet between Mars and Venus, we would have a carbon dioxide and water atmosphere and that would be it.'

5

**RICHES
OF THE EARTH**

If living on the Moon or in man-made settlements in orbit is conceivable, how much easier it should be to create decent living conditions on Earth! Just as James Lovelock looked back from Mars to perceive in a clearer light the living resources of the Earth, so we can compare the technologies of earthlings with those proposed for the space settlers. The hot deserts, mountain-tops, ice sheets and the vast unused surfaces of the oceans are all congenial compared with the surface of the Moon. If one imagines a terrestrial version of the Santa Claus Machine extracting any elements required from rock or sea water, the Earth is a treasure-house indeed.

Mining the water of the oceans *for compounds of magnesium, the Kaiser Refractories plant precipitates magnesium hydroxide by chemical action, in large tanks on the Californian coast. Sea water is already the world's principal source of magnesium metal, and of bromine too. These are tokens of the mineral riches available for exploitation if human beings choose an oceanic way of life, on the shore or on the oceans themselves.*

Human beings have, it is true, made large and potentially fateful inroads into the best of the metal ores and fossil fuels; thereby their economy is threatened with possible dislocation. By taking all the choicest land for farming and disdaining the living species and varieties that do not fit into their agricultural rosters, they endanger the one treasure unique to our planet: the genetic resources of life. But there are remedies that can restore human confidence in the riches of the Earth, and in its capacity simultaneously to meet the industrial needs of an advanced civilisation, to supply the food needed for a growing population, and still have room to spare for more abundant wildlife.

This aspect of our quest for big ideas was, in part, a re-exploration of possibilities about which I had speculated a decade before, in a book called *The Environment Game*. Then I had visualised a world which would combine advanced technology with a renewed regard for the natural landscape and the other species that share the planet with us. All parts of the Earth's surface, and not just the choicest bits, were in principle open to human habitation. Properly spread across the land and the sea, and condensed into settlements taking up little room, people could, I thought, give much of the Earth's territory back to wilderness – the management and enjoyment of which would become major themes of human life. With the passage of time this concept has grown more plausible, and practical ideas about how to do it have emerged from unexpected sources. It does not depend on artificial methods of making food.

An obvious scientist to consult about industrial prospects was Harrison Brown, professor both of geochemistry and of science and government at the California Institute of Technology. As a young research chemist in 1942 he was involved in the efforts to extract minute traces of the new element plutonium. After the war he helped to launch the modern study of the chemistry of the Solar System, by dividing all its contents into three classes: rocky, icy and gaseous. And in the early 1950s he took stock of the Earth's resources and the growing demands upon them – anticipating the concern about food, material and energy supplies that became fashionable much later. He went on to serve as foreign secretary of the US National Academy of Sciences, in a period when international collaboration in science was burgeoning, along with scientific perplexity about how to meet the needs of the growing human population. Reflecting on those needs, Brown came to the conclusion that industrial civilisation had passed a point of no return.

The numbers of people alive in the late twentieth century could not possibly be supported without a large input of science and technology. 'If you give it up,' Brown told us bluntly, 'a lot of people are going to die.' And if industrial civilisation were to disintegrate future generations might have great difficulty in getting it started again. The high-grade resources needed to lay the foundations would be lacking. People might cannibalise the remains of old technology for a while, but in general they would be stuck with a future very like the pre-eighteenth-century past. Brown considered such an outcome plausible, but not inevitable, for he held out the long-term prospect of a new and prosperous 'stone age' for industrial civilisation.

Provided technology and prosperity persist without interruption, there will be no limit to the lean grades of mineral resources that human beings will be able to use, until they are extracting minerals and fuels from the ordinary rocks of the Earth's crust. The rocks contain uranium and thorium, so that a ton of average granite incorporates nuclear fuel equivalent to 40–50 tons of coal. Of that, an equivalent of about fifteen tons will be isolated relatively easily from the rock, providing much more energy than will be required to extract it. The same granite will yield generous supplies of aluminium, iron, titanium, manganese and copper.

Brown was not urging anyone to invest in such an expensive process in the 1970s but he thought that, from a philosophical point of view, it was important to know that all the rocks of the Earth's crust could help perpetuate our civilisation by assuring it of a virtually unlimited supply of energy and materials. Given that confidence he wanted to see experiments in new ways of living on Earth and creating a comfortable life for everyone, before engaging in expensive space colonisation. Making industrial civilisation less vulnerable to damage and disruption was one objective that he suggested. It might be accomplished by building self-sufficient cities – industrial society's counterpart to peasant villages.

A cosmic gambler looking at the Earth would, Brown thought, place his bet against the survival of industrial civilisation. Seeing humanity split into two quite separate cultures, the rich and the poor, he might judge the rich to be the more vulnerable group. The traditional peasant culture is stable, in the sense that, if a few hundred Indian villages are wiped out by famine, the reproductive potential is sufficient to fill the empty spaces. When industrial societies make war with traditional societies, as in Vietnam, they see this resilience. By contrast, their vast interlocking network of mines,

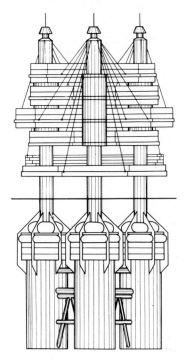

Kiyonori Kikutake, *an architect working in Tokyo, developed the ideas for ocean cities depicted on this page. He is a leading proponent of the idea that people should start moving their homes and industries on to the vast oceans, to relieve the pressures on the land. As a small-scale demonstration 'Aquapolis' (below) was built for an exhibition at Okinawa in 1975. The platform is 330 feet across.*

Floating cities *on large buoyant cylinders (diagram above), reaching hundreds of feet below sea level, would be almost immune from disturbance by ocean waves. The model (below) shows a scientific and recreational ocean city contemplated for Hawaii.*

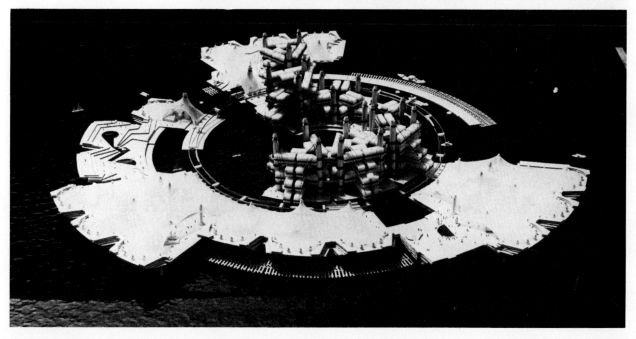

factories, communications systems, power grids and transport systems is fragile, and open to disruption by nuclear weapons, by an oil embargo, and by terrorism and sabotage.

The self-sufficient industrial city of Brown's imagining will be located near the coast, where sea water and ample rocks of various sorts are available. A population of 250,000 will be suitable. Energy will come from various sources ranging from waste plant material through solar heating to uranium and thorium extracted from the rocks; the rocks will also supply important structural metals. The oceans will be mined for other metals, while fresh water will come from salt water. The area surrounding the city will be farmed, with substantial use of greenhouses. The city itself will be designed compactly and people will not have to travel far to their work.

Such a city need not be on land at all: similar self-contained communities will stand on artificial islands or float in the deeper ocean. In Brown's opinion the floating ocean city can easily be made virtually self-sufficient – and much less expensive and more comfortable than an orbiting city in space will be. As visualised by the Japanese architect Kiyonori Kikutake, a structure a few miles wide will float on submerged concrete foundations that contain the machinery, allowing the creation of a clean and humane city above the water-line.

The chief products from sea water will be fresh water and large quantities of magnesium, a very useful lightweight metal. There will also be metallic nodules to mine from the floor of the ocean. As in the case of rocks, sea water will, if required, provide uranium as a nuclear fuel.

But in Brown's conspectus the ocean will offer its own special sources of energy, notably the possibility of using the difference in temperature between the warm surface water of the tropical ocean and the cold water that lies a few thousand feet underneath. The idea is to bring the cold water to the surface and generate power in plenty to supply a large ocean city. In the 1980s the US Energy Research and Development Administration will test a prototype power station using 'ocean thermal energy conversion'. And there will be an important bonus: the deep ocean water coming up will be rich in nutrients for marine life, and can be used to raise fish and shellfish in great abundance for food.

It seems to me that the oceans of the Earth have a rich endowment of energy and materials, judged even by the cosmic standards of those who would colonise the Solar System. They work as collectors of solar energy on a planetary scale, with the tropical surface water absorbing sunlight while cold water flows from the Arctic and Antarctic to maintain the great chilly reservoir underneath. Floating settlements that use ocean thermal energy and deep-water nutrients will be tapping two of the greatest unused resources of the Earth. They will make good the separation of nutrients from sunlight which makes the surface waters across wide stretches of ocean remarkably barren. Gaia seems perverse here: the natural system leaves phosphorus in notoriously short supply (in many parts of the land as well as the tropical ocean) while bottling up, in the dark water, deep materials equivalent to thousands of years' supply of agricultural fertilisers. Bringing up the water for power generation and fish-farming will mimic the natural upwelling of water off the coast of Peru which, albeit unreliably, sustains by far the biggest of all existing fisheries.

The large-scale colonisation of the oceans may well precede a large-scale migration to space settlements of the kind envisaged by Gerard O'Neill. Ocean living will multiply the areas and resources available to the human species, without the uncertainties and hazards attendant on the greater enterprise. Cities in remote parts of the ocean, aiming at independence and self-sufficiency, will be a useful dress rehearsal for many aspects of living in space. They will not, though, test the fundamental problems of biological viability discussed in the last chapter, because they will be bathed in the Earth's atmosphere and water.

In the hot deserts on land, abundant sunshine runs to waste for lack of water. Here, too, enlarging the territory of life on Earth will be a trivial task compared with carrying it to man-made settlements in space. Conceivably people will always say it is more economical to enhance the energy supply to well-watered places, than to take the water where the sunshine is. But that is an accountant's argument and the urge to make the desert bloom is as old as human civilisation.

Soilless cultivation *by the nutrient-film technique is a prime example of new methods that should enable our species to feed itself more reliably while making smaller demands on the Earth's land areas. The inventor is Allen Cooper, shown here at work at the Glasshouse Crops Research Institute in Sussex. The stems of the plants can be seen rising from simple channels where a circulating stream of water bathes their roots. The water contains all the nutrients needed for growth; these are topped up as the plants consume them.*

There will be two keys for unlocking the sunny treasure: one to supply water to the deserts and the other to make the best possible use of available water for growing crops. The latter may be the more important development but interesting possibilities for watering the desert abound.

Desert-dwellers will no doubt exploit more fully the large reserves of water in natural underground reservoirs: under the Sahara desert alone these will supply enough water to irrigate an area as large as England or New York State for 400 years, even if they are not replenished. By a scheme of the Indian Atomic Energy Commission, it would be possible to use a nuclear power plant to safeguard 6000 square miles of farmland in Uttar Pradesh against failures of the monsoon, by pumping up water from beneath the Ganges Plain and sharing the cost with manufacturing industry. But even these underground reserves will seem small compared with the fresh water locked up as ice in and around the polar regions, and in glaciers elsewhere – a hundred times more fresh water than in all the rivers and lakes of the world. To melt the ice sheets and glaciers on land might tend to raise the worldwide sea level, however slightly, but taking ice in the form of icebergs and floes already afloat will not have that effect, and it will in any case be readier to transport. If Gerard O'Neill can reckon on slinging millions of tons of material into space, moving ice about on Earth ought not to seem an extravagant aim.

Towing icebergs may become a routine activity in the southern oceans. The icebergs spawned by Antarctica each year could in principle supply about 200 tons of fresh water for every person on the planet: in practice, only a small fraction of that potential is likely to be trapped and delivered to regions most in need of water, notably Australia, the Middle East, Africa, Chile and California. There will remain, of course, the open-ended possibility of obtaining as much fresh water as people want, by removing the salt from sea water. For ocean dwellers and the inhabitants of Harrison Brown's self-sufficient coastal cities, who are processing sea water for other purposes, that will be the natural thing to do. But for everyday use on land, especially for agriculture, desalinated water may always tend to be expensive because of the energy needed to produce it.

Strange to say, the task of making the deserts bloom with limited supplies of water will become easier when plants are cultivated in water. Traditionally, farmers have watered the soil and let the crops extract moisture as needed, but it was always a wasteful way of doing it, with most of the water running off or evaporating without passing into the plants. When it becomes the general practice to feed the water directly to the roots of plants, far greater production of crops will be possible for a given supply of water. Scarcely more water will be needed than the plants actively employ in transpiration. Given the necessary minimal supplies of water to match the sunshine, deserts can in principle become major centres of crop production.

The idea of doing away with soil and growing traditional crops in water instead – so-called hydroponics – aroused a lot of interest in the 1930s but nothing very much came of it except for experiments with artificial soil. In Britain in the 1960s, Allen Cooper of the Glasshouse Crops Research Institute pioneered a route to true hydroponics. In his 'nutrient-film technique', the plants grow in gently sloping gullies made, for example, of plastics. A shallow stream of water, enriched with the appropriate nutrients and trace elements essential for life, flows over the roots and recirculates. The plant roots extract the water and chemical ingredients that they need to grow and sustain life. The roots also mesh together in strong mats which support the plants.

A greenhouse that I visited, equipped for this technique, had chemical monitoring instruments and automatic-metering pumps for topping up the nutrients in the circulating water. It was fascinating to listen to the machinery speeding up when the sun broke through the clouds. You could *hear* the plants demanding more nutrients in the sunshine. Much as a conventional greenhouse controls the air around the leaves, the nutrient-film technique controls the environment of the roots in a very precise way. But although most of the early trials were done by greenhouse operators already accustomed to regulating the environment of their crops, the technique also works in the open air.

One great advantage of using soil in the 10,000-year-old way was that farmers could make mistakes and yet their crops would tend to survive. Soil is a powerful regulator of the chemical environment of the roots. Mistakes with the nutrient-film technique can, and do, kill crops. As a new method of cultivation, it will require extensive research and experience before the innovators will know its full scope and limitations. Even if all goes well, care, vigilance and skill of a somewhat higher order than usual will be needed, as growers in many countries adopt the technique.

Although the pioneers of the nutrient-film technique have been judiciously cautious in their published expectations, I suspect that its full potential is astounding. It is a big idea which may revolutionise crop production all over the world. Viewed in an ecological way it offers many interesting possibilities. When fish-farming and nutrient-film culture are combined, the fish will enrich the water with nutrients, which will then serve for raising crops. Human sewage effluent may serve as a nutrient medium for growing grass, and so help purify the water at the same time. And because the system works even when the nutrients flowing over the roots are dilute, river water will, in some parts of the world, be able to support highly productive nutrient-film culture of cash crops and food along their banks, without any need to add artificial fertilisers. Growers in industrialised countries can automate the system and, when workers are fed up with stooping, the channels can be raised to a comfortable table-height. Alternatively, in poor countries with high unemployment, many people will be able to busy themselves with caring for crops and maintaining the nutrient supply at the right level. This flexibility in the manner of operation will be one of the attractions.

If the promise of the nutrient-film technique comes true, the big pay-off will be in raising crops without soil and needing only a bare minimum of fresh water – starting with the hot deserts. But people will also be able to use it on rocky ground, on mountains, on the roofs and walls of cities, on ships, on artificial islands and in space settlements. Anywhere that has daylight to make crops grow will become a potential site of food production.

According to some people, you won't even need daylight. Eion Scott and his 'controlled environment agriculture' group in General Electric at Syracuse, New York, tested the possibility of growing plants by artificial light, in a system that incorporated the nutrient-film technique. They considered that winter vegetables grown in New York State by artificial light would become competitive with vegetables transported from California or Mexico. In the completely controlled environment, they produced an annual yield ten times greater than a greenhouse's and twenty or fifty times greater than a field's. For economy in water, the GE system combined air conditioning with recovery of the water transpired by the crops. As a result the water consumed per pound of produce was less than a tenth of the consumption in a field. Even after powering the lamps, the energy consumption could be less than in a conventional greenhouse.

Given a suitable supply of energy – which is of course the big question – systems of such a kind could in principle allow people to live anywhere. They could for example settle in Antarctica and enjoy fresh food through the long polar night; they could even produce

food underground. Scott sought to engage the interest of the US space administration in this development, offering it as a basis for experiments in food production in space. But the ideas of other scientists ran in quite the opposite direction – towards making the best possible direct use of the Sun's energy.

A vivid picture of what even conventional farming techniques could achieve came in 1970 from a Dutch plant physiologist, E. C. Wassinck of the Agricultural University at Wageningen. He pointed out that primary yields no greater than those prevalent in good agricultural practice (converting one per cent of the energy of sunlight into photosynthetic products) could comfortably support the entire human population from a farmed area of only three per cent of the land of the Earth. That was after making quite generous allowances for cattle raising, wood production and so on, and for roads, buildings and water supplies needed by the farmers. Wassinck's idea was to collapse the world's agriculture into about 3000 units, 'scientific and industrial enterprises', each covering an area about twenty miles square, and suitably scattered around the world.

Thus Wassinck wanted to make the individual farm much bigger, while 'decultivating' large areas of the Earth. He thought that concentrating food production into large enterprises was not much more strange than concentrating car production, and he argued that the impact on the natural environment would be less from one large area than from a corresponding area of small units. Yet by basing his argument on conventional field agriculture he was understating the possibilities of compact food production and asking for more land than may really be needed. And the people pursuing those other possibilities tended to favour smallness and self-sufficiency rather than huge agricultural enterprises.

On Prince Edward Island in the estuary of the St Lawrence River in Canada, a new concept of living and growing food was being put to the test in the late 1970s. Called the Ark, it was a world in miniature – a self-contained system for supplying food and energy from a compact area. The climate was often harsh outside, but inside the Ark it remained tropical. The Ark was the brainchild of John Todd and his New Alchemy Institute, which had earlier created a similar system in Massachusetts. Todd's aim was to demonstrate how a minimum of inexpensive hardware, that any family could understand and run, would reliably keep people alive and prosperous whatever the state of the global economy.

Standing on a headland overlooking the river that froze each winter, the Ark on Prince Edward Island was

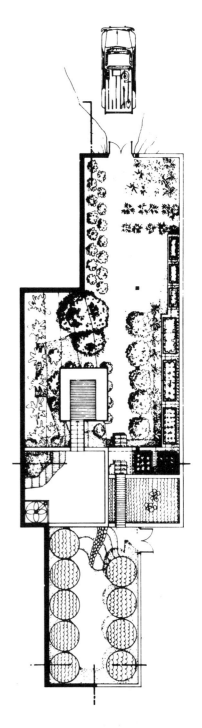

The 'Arks' of the New Alchemy Institute *fulfil the ideas of John Todd* (left). *Family-sized greenhouse systems provide a surplus of food. They rely on solar and wind energy and apply ecological principles in the production of vegetables and fish. The plan (below right) shows the Cape Cod Ark in Massachusetts, where concepts are tested; the circles are fish tanks. The sections (below left) are of the Prince Edward Island Ark in Canada. The first is the barn end, with the main greenhouse and fish tanks plainly showing; under the barn is a volume of rock for heat-storage. The lower section shows a household built into the Ark.*

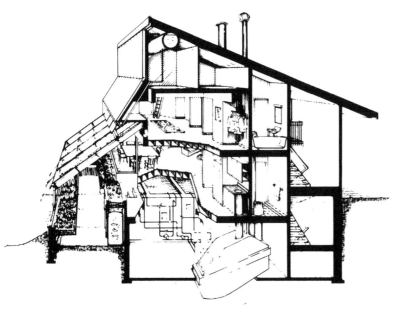

a fertile fortress against the elements. It was conceived as a three-bedroomed family house wrapped in its own food-producing and energy-handling systems, which trebled its size. Trees and an earthen bank to the north protected it against the elements. From the north, the Ark looked like a fairly conventional house, but the south side was translucent, to draw in the sunshine.

The Ark was in essence a combination of greenhouse and fish farm, but behind the apparent simplicity lay a lot of fresh science and technology. Panels rising above the greenhouse were passive machines for trapping solar energy as effectively as possible. Besides growing vegetables for domestic use and for sale all the year round, the Ark raised several crops of fish in so-called solar ponds, which also stored heat. As a former student of fish behaviour, Todd was interested in substances secreted by fish which encouraged docile behaviour in a dense community. It turned out that these fatty substances helped to reduce evaporation from the surface of the water and also made it especially good for watering the vegetables. The ecological design was based on some extremely productive natural ecosystems in rivers in Java and the southern United States. The aim was reliability rather than record yields, but in fact the yields turned out to be the highest ever achieved in ponds.

Todd complained that when he was an agricultural student in Montreal he was taught about the industry and business, not about the care of living systems. So he switched to studying the biology of behaviour, later returning to agriculture with the resolve to put the biology back into it. He also wanted to combat the deadly destruction of biological and human diversity on our planet, which was going on in the name of efficiency and profit. The Mexican dwarf wheat of the 'Green Revolution', for instance, was a remarkable product of the plant breeders' art, but it caused the extinction of hundreds of local varieties of wheat in the countries where it was introduced. Todd feared that the world was heading into a trap like that sprung upon the Irish in the 1840s, when the blight struck their newly specialised fields of potatoes.

When Todd and his colleagues were looking for pest-resistant vegetables for the first Ark, in Massachusetts, they found that modern plant breeding had neglected that natural quality, presumably because breeders took for granted a universal and perpetual use of chemical pesticides. Renouncing pesticides, the New Alchemists experimented with insect-repellent herbs and flowers, and also found lizards to be very effective eaters of insects.

The Arks relied exclusively on sunlight and wind power, with collection panels for solar heat supplying domestic hot water and storing heat for the buildings as protection against a long cold spell. When sunlight failed, electricity from the wind kept them warm; it also ran the electrical equipment. The windmill used in Canada was of an advanced design which transferred energy by a hydraulic system to a generator on the ground.

The early experiments were necessarily small — perhaps too small, too family-like and peasant-like for many people's tastes. But there was nothing primitive about the information and control systems and Todd considered the scale none too small for the immediate practical needs of the world's rural poor. Communities like the Arks could be set up almost anywhere on Earth, provided the engineering and biological design was re-worked to suit local climates and life-styles. But in warmer regions like Central America, where a new project was getting under way, the system would run more easily than on chilly Prince Edward Island.

The Arks were a way of life, which the New Alchemists offered as a model along with their integrated system of modest technology. They were egalitarian: all members of the staff had the same salaries and decision-making was by consensus. Although their institute was a scientific one they deliberately 'diluted' the staff with non-scientists, who did much of the research. In Todd's view, to tend and test a living system came naturally to all humans and the cultivation of 'lay science' was an important aim in its own right.

The name of the New Alchemy Institute was not frivolously chosen; it reflected a belief in the unity of knowledge, in the manner of the old alchemists. The secrets of Hermes, by which the alchemists organised their thoughts and actions, told of a unity of nature, pervading Earth and sky, whence all things came by adaptation. Todd thought it translated neatly enough into modern cosmology and evolution theory. And if the twentieth-century alchemists spoke of feedback systems instead of fire versus water, and sought ecological principles instead of the philosopher's stone, they shared the goal of linking science and scholarship, practical action and philosophy, in a comprehensible whole. As the world in microcosm, the Arks could not be broken up, or bought and sold.

Thus the Arks represent another big idea. As such its future will, I think, depend on whether people want to live that way. Those who do may be able to sidestep the powerful economic arguments that favour open-field agriculture on cheap land, by attaching a high value to

Night-time insulation *for a greenhouse roof is provided by detergent foam (upper photograph) in experiments at the University of Arizona. Such techniques may* enable *communities to grow cereal crops in large greenhouses (model,* below*) according to the proposals of Theodore Taylor.*

self-sufficiency and reliability in food supplies. Similar considerations surround another approach to compact food production, possibly complementary to Todd's, pursued by Theodore Taylor. I introduced him as begetter of the concept of the Santa Claus Machine (see Chapter 1). In explicit contrast with his ideas for large-scale operations in space Taylor favoured the small scale on Earth. And the practical question he addressed was: Why not grow wheat under glass?

The traditional greenhouse is an expensive way of producing highly-priced vegetables and ornamental crops, but it gives the advantage of high and reliable yields, almost independently of the vagaries of weather and climate. Taylor urged that greenhouses ought to be made much cheaper to build and run, so that all crops, including mankind's staple cereals, might be produced under their protection. For instance, environmental engineers at the University of Arizona tested a trick for keeping the greenhouses warm in the winter. At night they fed a liquid foam between two layers that formed the roof, to serve as an insulating material, and removed the foam during the daytime. As expected, the foam greatly reduced the heat losses.

Available methods of solar-energy conversion will, in Taylor's view, eliminate the need for extraneous sources of power for greenhouses. For ventilation, electric fans will be replaced by solar chimneys. Coloured black at the base to absorb sunlight, these will expel the air by warming it, thus drawing in other, fresh air with the supply of carbon dioxide needed by the growing plants. And mirrors can focus sunlight to raise steam, which will then react with organic waste material from the greenhouses to make hydrogen – a convenient fuel for many purposes. By a sensible combination of such techniques, in greenhouses of modular design say 800 feet long by 120 feet wide, it will be possible by Taylor's reckoning to grow wheat in greenhouses for as little as twice the prevailing cost of growing wheat on a Kansas farm.

Several advantages will offset that extra cost. Wheat and all other food will be grown where it is eaten, thus cutting out the long chain of distribution to the consumer. There will be no wastage, nor any need for processing the food or adding chemicals to preserve it. Fertilisers and pesticides will be kept under control within the confines of the greenhouse economy. Sunlight, water and air will come into the greenhouse, harvested food will come out, and almost everything else will be recycled.

In Taylor's vision, small communities anywhere in the world will produce all of their food within walking distance. In the United States, for instance, the total added cost to a community of the greenhouses to supply 'a good, balanced, more or less traditional American diet' should be less than a tenth of the cost of building the associated houses, roads and so on. By adding other solar-energy collection and storage facilities, connected as appropriate with the greenhouses, there will be near self-sufficiency in energy as well as food. In a 'globe of villages', as opposed to a global village, political control will be as thoroughly decentralised as food and energy production.

Decisions affecting their lives on a day-to-day basis will be made by the people in each community. For Taylor the ideal number for a village community, or the self-sufficient neighbourhood of a city, will be 2000–4000 – the number of people who can know each other, at least casually. The technology of the late twentieth century will thus be a way of fulfilling Mahatma Gandhi's dream of the village culture, creating a world in which everyone can be important. Military tensions will be reduced, because self-sufficient communities without any special resources will be hardly worth attacking. There will still be scope, in Taylor's world, for people to join forces on a big scale, to explore the moons of Jupiter and Saturn, for example. But the first order of business will be to bring government and industry 'right side up' and retrieve local self-determination.

It is important, I think, to tease apart the technical and the political strands of ideas such as Todd's and Taylor's. I am not denying that the two things often run together both in conception and in practice, and I was certainly interested to find among scientists and technologists this current of thought flowing strongly in the direction of material and political independence for small groups of people. It eddied through the writings of Gerard O'Neill and others in praise of space settlements, while the late 'Fritz' Schumacher gave it a slogan: 'Small is beautiful'. But recall Harrison Brown's picture of the self-contained industrial city of 250,000 people, and Wassinck's superfarm serving more than a million people. These are on the agenda of possibilities too. I suspect there will always be a full range of human settlements, on Earth or in space, from isolated family units to the giant cities which many people actually like.

More significant, in my opinion, are the technological trends which promise to alter the relationship of human beings to their planet. We have sampled a range of ideas, prevalent in the closing decades of the twentieth century, pointing to human life conducted in ways that will reduce the human demand for land. They will enable people to establish their communities anywhere on the Earth's surface. The occupation of oceans and deserts will moderate the pressure of populations in other places. And the compact methods of food production should in principle liberate land. If people contrive to live, work and produce their food in much smaller and more concentrated areas, the way will be open to a restoration of large parts of the Earth's surface to something approximating to its pristine glory.

Giving the land back to the trees and wild animals will reverse the process whereby the farmers took all the choicest soil and best-watered land of the world and decked them with their peculiar and arbitrary selection of plants and animals. Any encroachment by the natural diversity of living things was a sign of slovenliness. Starting in the Neolithic period when the first of them attacked the forests of Europe with flint-axes and fire, the farmers destroyed the natural environment on a wider scale than miners or manufacturers ever attempted. The basic reason was their inefficient use of sunlight, which obliged them to collect it over vast areas. When people come to reconstruct the pre-agricultural landscapes, they may turn first to the palaeobotanists' pollen charts to see how nature seeded each region before the farmers came. Pollen grains identified in layers of ancient soil provide lucid records of the changing patterns of vegetation at a given place. They show the impact of human activity and reveal many cases where supposed 'wilderness' is the unnatural consequence of human action.

For authoritative ideas about the future relationship of human beings and wildlife, I sought out Edward Wilson at Harvard. Wilson sprang into the public eye in the mid-1970s as the founding father of sociobiology, setting out to find evolutionary reasons for the way animals – including humans – behave in social groups. But I had known him before that simply as an outstanding zoologist – an expert on the social insects and one of the disturbingly small band of biologists who attempted to put a scientific backbone into that collection of vague generalisations and parochial details which passed for ecology.

We began by talking of the conjectured move into space. Wilson said that the paucity of species to be expected in a space settlement was one of the more depressing aspects of such proposals – and he insisted that he was not just expressing the prejudice of a zoologist. The great interest that people in industrial societies showed in the living diversity around them, and the care they put into 'non-functional' pets and flower gardens, were for Wilson signs that a love of other creatures and their diversity was a deeply-rooted biological and psychological trait in human beings. He suspected that it stemmed from the days when our ancestors needed detailed knowledge of and concern for animals and plants, just to survive. The consequences of living over a period of generations with the diversity of other living things drastically reduced might be deprivation or brutalisation – at any rate some important change in the human psyche. By Wilson's prescription, space settlements ought to include nature reserves, surrounding the people with as much diversity of plant and animal life as may be technically possible, in order to provide them and the generations ahead with this special psychological nourishment.

On Earth, given that human trait, the care and study of the literally countless species that share this planet will, in Wilson's judgment, keep a substantial segment of mankind busy for generations to come. It will absorb the energy and emotions that made us good at hunting and farming, and people with the instincts of bird watchers, butterfly collectors and backyard gardeners will be able to sustain their interests indefinitely as ecologists and biogeographers. The immediate task will be to halt the reduction of organic diversity brought about by careless and unmonitored modification of the environment. Setting aside a few isolated nature reserves has been an inadequate policy, causing a decline in the number of species supposedly under protection.

The reasons for such a decline became well understood as a theory of biogeography began to develop in the 1960s and 1970s. Wilson himself was involved in an experiment off the coast of Florida, designed to test the proposition that the number of species that a given area could support was pretty well fixed. Very small islands, mere clumps of mangroves inhabited by insects, spiders and other small animals, were covered with tents and fumigated. Two years later, other insects and spiders, often of species quite different from the original inhabitants, had recolonised the mini-islands, but the numbers of species were approximately the same as before.

Continuing studies of that kind can become a guide to how human beings will best maximise the diversity of plants and animals on Earth. And the design of nature reserves will, by Wilson's reasoning, be a matter of geometry. One large reserve will save more species from extinction than several small ones of equivalent area, and a circular reserve will be better than an elongated one. When small reserves are established to save special habitats, they will be clumped as closely together as possible and linked by natural corridors; a given species will be unlikely to die out in all the

Geometry of nature reserves most likely to ensure survival of species (after Wilson).

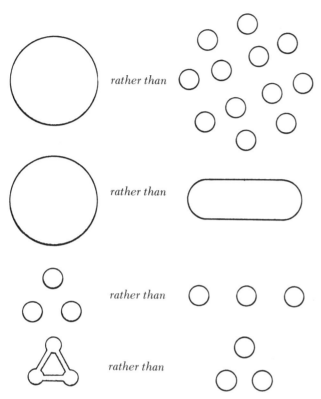

rather than

rather than

rather than

rather than

Edward Wilson *of Harvard has thought more deeply than most other biologists about the relationships of nature and human nature. As a founder of sociobiology (see Chapter 10) he considers that many features of human emotional and social behaviour have their origins in evolution, and will not easily be modified. One such feature is our love of animals and other wild life, and Wilson thinks that space settlements should include nature reserves to supply a necessary form of psychological nourishment. He also sees the creation of new species of plants and animals as a long-term goal for mankind, in keeping with what he calls 'the ethic of diversity'.*

On the other hand, as an investigator of the ecological species-capacity of small areas, Wilson thinks that policies for conserving existing wild species on the Earth are inadequate. They do not take sufficient account of what is known about the geometry of survival (diagrams, above). A network of large reserves, connected by wild corridors, is more likely to prevent the annihilation of species than are small, isolated reserves like present-day national parks.

reserves at the same time, and the survivors will be able to recolonise another area provided they can reach it.

Entirely new communities of life will also be created. A small token of the possibilities came from marine biologists in Florida and California, who dumped rubble, old cars and other refuse in shallow water. Thus they created artificial reefs, on which many different organisms quickly established themselves. The challenge will be greatest in those great areas of the Earth which nature populated only very sparsely with indigenous plants and animals: the deep oceans, the ice sheets, the dry deserts, and tracts of land like those in Australia where a lack of crucial trace elements in the soil hampers life. Human beings will be able to establish communities of wild species in combinations that nature never tried.

In these activities, humans will already be playing God with the species. They will be preserving species that were due for extinction for natural as well as man-made reasons. They will be transplanting species from other parts of the world, more judiciously than in the past, in ways that will actually benefit the local species and perhaps save the transplanted species from extinction. Within the reserves, by cataloguing, census-taking and monitoring of all of the plants, animals and microbes living there, people will arrive at much more sophisticated knowledge of ecology and the process of evolution. When they have come closer to solving the 'species-packing problem', to determining how sets of different species can be combined for maximum diversity, they will be ready for the next big step.

It will be the creation of new species of plants and animals. That process may take generations or even thousands of years, but it will be within our power to accomplish it. The motive, according to Wilson, will be 'the ethic of diversity' and the goal will be to surround ourselves with a rich variety of genetic types and species. We shall become managers of living nature in a much closer manner than ever before, using an ever more precise and humane science of ecology, and 'biogeographic technology'.

When the creation of new species is to be the aim, I should add that the local strategy of reserve allocation may be rather different. One of the most powerful ways in which nature creates species is by separating small groups of animals from the general population of their kin. The rifting of continents, the building of mountains, the encroachment of deserts, the eruption of volcanoes creating new islands – these are among nature's agen-

cies. The unprecedented number of species inhabiting the Earth in our geological period may be due primarily to the break-up of the former supercontinent of Pangaea. The deliberate evolution of new species may require small, isolated areas to be set aside for the purpose, rather than the large or connected reserves prescribed by Wilson for the preservation of species.

Genetic engineering will also have a part to play in the creation of new species, becoming, in Freeman Dyson's phrase, 'a new art form'. A modest start has been made with the new nitrogen-fixing bacteria mentioned earlier. For higher organisms there are tricky matters about the control of gene function, yet to be understood. But Dyson himself speculated about re-engineering living plants so that they will make glassy materials and grow their own greenhouses around themselves. That will be to repeat in the plant kingdom an invention which nature contrived a long time ago in the animal kingdom – 'creating the first warm-blooded plants', as Dyson put it.

Meanwhile our giraffes and geraniums are the only giraffes and geraniums in the Solar System, or in the Galaxy. In that sense they are to be counted as cosmic riches, and people will need to take many such species into space with them. There will be no conflict of purpose between making the Earth an even better place for life, and making the Solar System habitable.

In the opinion of James Lovelock, author of the Gaia hypothesis (see Chapter 4), we have to puzzle out what the role of our species is to be, in the task we have in common with all other species of maintaining the Earth as a planet fit for life. One of the ways in which human beings will serve Gaia, in Lovelock's opinion, will be by providing a brain and nervous system for the planet. Our global telecommunications network operating at the speed of light, and the satellites that keep the Earth under constant surveillance, are already tokens of that function.

As its brain and nervous system, human society may also defend the living planet against cosmic disasters. Consider the danger of collision with asteroids: when these hit the Earth in the past they did not kill all life but the larger ones caused great devastation. As Lovelock pointed out, it will be possible to avert such catastrophes in the future. When astronomers compute that the orbit of a massive-looking body is intersecting the Earth, the most powerful rockets available will intercept it, carrying H-bombs. Exploding to one side of the asteroid, the bombs will knock it on to a safer course.

Meteor Crater *in Arizona* (above) *was made by the collision of a very small iron asteroid with the Earth. The asteroid had an estimated mass of 65,000 tons and released on impact energy equivalent to a two-megaton H-bomb. Far bigger asteroids orbit the Sun close to the Earth and threaten to hit our planet during the next few million years. Eugene Shoemaker* (right), *who pioneered 'astrogeology' for NASA, now nurtures an interest in the Earth-approaching asteroids as the next target for manned exploration.*

6

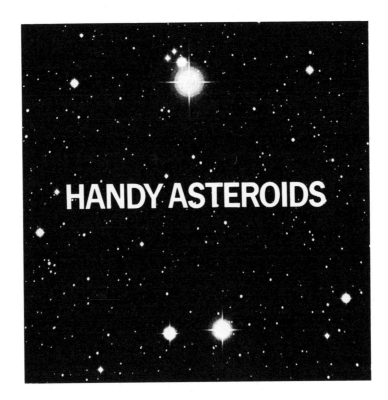

HANDY ASTEROIDS

The diminutive planets swarming around the Sun gripped the imagination of two people who, by coincidence, had both quit the US space programme in a state of disillusionment about the official attitudes to science. At the California Institute of Technology, Eugene Shoemaker told us he wanted to see an asteroid as the next cosmic target for manned exploration. He had founded the 'astrogeology' branch of the US Geological Survey, sponsored by the National Aeronautics and Space Administration. He was responsible for planning the geological work of the *Apollo* venture and helping to train the first astronauts to visit the Moon. But he found he could not get NASA to take the geology of the Moon as seriously as he wished. When there seemed to be no plans even to send a geologist-astronaut there, he gave up the work before the first *Apollo* landing and went to Caltech as professor of geology.

The other asteroid enthusiast was Brian O'Leary, who prophesied their use in building space settlements and satellite solar power stations. As an astronomer, he was selected for training to be a scientist-astronaut but gave that up and wrote a book: *The Making of an Ex-*

Astronaut. I had first met him briefly in 1968 at Cornell University where he was expounding the case for putting telescopes into Earth orbit, to study the planets. But a decade later I found him at Princeton, as Gerard O'Neill's closest colleague in the conception of space settlements. While O'Neill was emphasising the mining of the Moon, O'Leary was suggesting that asteroids might be a better source of raw materials in space. This chapter describes the potential importance of the asteroids for human action in the Solar System, mainly as foreseen by Shoemaker and O'Leary.

Shoemaker's interest in asteroids began when he studied the famous Meteor Crater, 4000 feet across, in the desert of northern Arizona. It was made by an asteroid hitting the Earth. How often, he wondered, did collisions of that magnitude occur? The answer turned out to be about once every 25,000 years, while bigger and more devastating impacts would occur less frequently. After quitting *Apollo*, Shoemaker's interest shifted to the asteroids that were approaching but missing the Earth. He saw them as the easiest places for astronauts to reach, after the Moon. He thought there would be about a thousand of them large enough to detect with a modest telescope – Earth-approaching asteroids about 2000 feet in diameter. Of those, about twenty might be very convenient for visits by manned spacecraft in this century.

The first asteroid was discovered in 1801, orbiting around the Sun beyond Mars. Thereafter many others were detected between the orbits of Mars and Jupiter, scattered across a huge ring of space from $2\frac{1}{2}$ to $5\frac{1}{2}$ times the Earth's distance from the Sun. The largest of them, Ceres, turned out to be about 630 miles in diameter, with a surface area about as large as India. A wide range of descending sizes was inferred – mountain-size, iceberg-size, and so on. Most of these lumps of rock and other material had presumably been circulating since the Solar System began. An idea that there might also be a lot of fine-grained dust in the asteroid belt was dispelled by *Pioneer* spacecraft passing through on their way to Jupiter.

The Earth-approaching asteroids were not necessarily typical. Shoemaker considered them to be mostly the cores of worn-out comets, perhaps a mile or a few miles in diameter. Comets themselves originated from a giant cloud of small icy objects that orbit the Sun at very great distances – too far to be visible from Earth. Occasionally, so the theory went, the objects were perturbed by the passage of nearby stars, and fell in towards the Sun. Most of them would just streak through the Solar System and out again, but a few, passing close

to the giant planet Jupiter, were trapped into orbits that took them around the Sun frequently. A few of those, in turn, fell under the influence of the inner planets and adopted even tighter orbits. The ices evaporated, leaving a stony core.

'Practically nineteenth-century science' was how Shoemaker described the methods being used at Caltech in 1977, to look for these handy asteroids. Certainly the effort seemed small indeed, at a time when people were talking of great treasure to be had from these objects. Shoemaker's colleague Eleanor Helin and her students would go up to Palomar mountain every month for four or five nights and seek out asteroids passing close to the Earth, by means of an 18-inch telescope, the famous observatory's smallest instrument. They would photograph a selected area of the sky directly away from the Sun for an exposure of twenty minutes while the telescope tracked the stars. Then they would scan the plate for short streaks of light – the signatures of objects such as asteroids that were moving relative to the stars. A second, shorter exposure taken after an interval confirmed that the astronomers were not just seeing a flaw in the first plate, and also revealed the object's direction of motion. When Helin and her students detected an object in that way they would follow it on successive nights and compute its orbit. Occasionally it would turn out to be an unpublicised man-made satellite of unknown purpose. Sometimes a barren period of four or five months would pass with no discoveries of natural asteroids.

'If we had larger optics,' Helin commented, 'we could increase their detection and discoverability by several orders of magnitude.' Even so, she had her successes and was discovering Earth-approaching asteroids at a rate of about three a year. In 1976, for instance, she found two asteroids in orbits very similar to the Earth's and coming very close: 1976AA and 1976UA. The latter passed within half a million miles of the Earth. Both of these objects, like the asteroid Hermes which in 1937 approached even closer, are liable eventually to crash into the Earth.

The theory that the nearby asteroids were remains of former comets gave Shoemaker and fellow astrogeologists a sharp scientific reason for hoping that the asteroids might be visited. Samples from them would provide important clues to what was happening when solid bodies first appeared during the formation of the Sun and the Solar System, and hence would lead to a better understanding of how the Earth was born. In what sequence and in what regions of the nebula of gas and dust around the Sun were the pieces of the comets

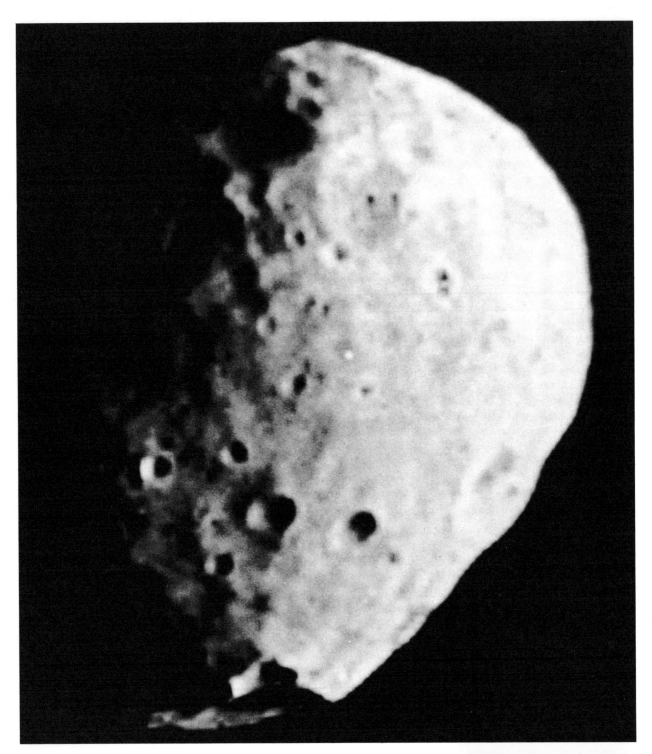

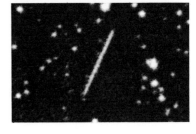

Phobos, *moon of Mars, is thought to be a captured asteroid and gives an impression of what asteroids in general probably look like. This picture, transmitted by the* Viking 1 *orbiter operating in the vicinity of Mars, shows many craters in the surface of Phobos, caused by collisions with lesser asteroids. The streak (right) was produced by an asteroid during a prolonged photographic exposure in a telescope tracking with the 'fixed' stars. This photograph, obtained at Palomar by a team from Caltech, shows a newly discovered asteroid, 1976 AA, which comes unusually close to the Earth.*

put together? Were they primeval building-blocks of the Earth and other planets? Careful work by a visiting astronaut, on a one-year round trip, would give invaluable answers to such questions.

Meanwhile astronomers elsewhere were scrutinising the colour, brightness and other features of the light reflected by the asteroids. In the mid-1970s they were reporting that most asteroids looked very like meteorites. That was not surprising, because the stones and lumps of metal that fell out of the sky to be picked up as meteorites were supposed to originate from among the nearer asteroids. They were the only pieces of the Solar System available for direct study by scientists until the astronauts brought back samples from the Moon.

The asteroids examined by telescope seemed to resemble one or other of the three main classes of meteorites – 'carbonaceous', 'silicaceous' and 'iron-nickel'. You could infer the existence of important material resources in the asteroids, from the chemical analyses of meteorites found on Earth. Water, which was non-existent in the samples from the Moon, would be available in the asteroids, judging by the carbonaceous meteorites which contained anything up to twenty per cent of water molecules incorporated in their crystals. As their name implied, carbonaceous meteorites also had a far more generous endowment of carbon than the Moon, much of it in the form of hydrocarbons. The fact that carbonaceous meteorites were scarce among the harvests of meteorites at the Earth's surface was not a sign that they were in short supply; it was a consequence of the burnable, crumbly and rapidly weathered nature of the material. In the Solar System, carbonaceous asteroids were extremely common, making up most of the asteroid population in some zones. The biggest asteroid, Ceres, seemed to be carbonaceous.

The silicaceous meteorites, and therefore presumably the silicaceous asteroids, scored in a different respect. They possessed an abundance of free iron – that is to say, in ready-made metallic form rather than in chemical combination in an ore. About ten per cent of the weight of typical silicaceous meteorites was iron metal, mixed with nickel. And on Earth some meteorites turned up consisting almost entirely of iron-nickel; for instance the meteorite that made Meteor Crater was a single lump of iron-nickel. Similarly some asteroids, judged by their light, were mainly metal. But the same asteroid was unlikely to contain abundant free metal as well as carbon and water.

Being, by definition, small objects that dashed across the sky, asteroids were hard to observe in detail. Many of them winked noticeably, implying that they had irregular shapes that were turning as they travelled, presenting now a larger, now a smaller face to the Earth. But two objects in close orbit about the planet Mars were captured asteroids – or so astronomers suspected. These little moons of Mars, called Phobos and Deimos, had been a matter of speculation for almost a hundred years; indeed the distinguished Russian astrophysicist, Joseph Shklovsky, suggested in 1960 that Phobos might be an artificial satellite launched by an extinct race of martians. But *Mariner* and *Viking Orbiter* spacecraft visiting Mars in the 1970s took close-up pictures of them. Carl Sagan of Cornell, at whose urging the *Mariner* 9 camera was first turned on Phobos, commented: 'Phobos looks not so much like an artificial satellite as a diseased potato.' The captured-asteroid theory was preferred, and the martian moons gave a special opportunity for seeing what an asteroid might look like at close quarters.

Both Phobos and Deimos turned out to be dark grey in colour and irregular, vaguely ellipsoidal bodies, their longest diameters being seventeen miles for Phobos and nine miles for Deimos. They bore the craters produced by impacts of smaller asteroids: two large craters on Phobos were named Voltaire and Swift, in honour of authors who wrote about imaginary moons of Mars. The high rate at which Phobos warmed up each time it emerged into sunshine from the shadow of Mars suggested that it might have a dusty surface – but perhaps only because of its special situation near Mars. Phobos was detectably losing orbital energy (it will crash to the martian surface in about 100 million years' time) and curious grooves seen in the *Viking Orbiter* pictures of Phobos were interpreted as cracks produced by the increasing tidal force as Phobos inched closer to Mars. Judging by its own weak gravitational influence on the *Viking* spacecraft as it passed close by, Phobos seemed to be a low-density, carbonaceous, water-rich asteroid.

On the basis of the advancing knowledge of asteroids and the inferences from the meteorites and the moons of Mars, Brian O'Leary reviewed the possibilities of mining raw materials from asteroids to support human life and construction in space. By 1977 he was judging the prospects to be very bright. Reckoning on the existence of an estimated 100,000 near-Earth asteroids larger than 300 feet in diameter, he thought some of them would turn out to be decidedly cheaper to quarry than the lunar surface.

The asteroids will provide plenty of iron and nickel, much of it in metallic form, but less aluminium and titanium than in Moon rock. The most important benefit will be the water, present in abundance in some asteroids but apparently missing on the Moon. That water, together with carbon and nitrogen, will help to make space settlers more independent of supplies from the Earth. Besides chemical composition and proximity to the Earth, one of the factors affecting the choice of a target asteroid will be the angle at which the asteroid's orbit slants, relative to the Earth's orbit; the smaller the angle the better. Modest-sized asteroids could be captured whole and installed as additional moons for the Earth, the first perhaps by the late 1980s. (O'Leary's timescales resemble O'Neill's in their audacity.)

Fifty flights of the Space Shuttle will be needed to fulfil O'Leary's scenario for lassoing an asteroid with an unmanned spacecraft. The Shuttles' loads will include the pieces of two electromagnetic 'mass drivers' (see Chapter 3) to be assembled in orbit about the Earth: one to send the spacecraft on its way to the asteroid, the other as the main component of the spacecraft – the solar-powered engine for bringing the asteroid back. When it reaches the asteroid, the automated spacecraft will sidle up to it and latch on to it. Using equipment described by O'Leary as 'essentially a garbage compacter and drill' the spacecraft will make slugs for its mass driver from the body of the asteroid itself. Then the mass driver, feeding on the asteroid, will slowly nudge it in such a way as to bring it into orbit about the Earth. Most of the mass of the asteroid will be sacrificed: for example, a ten-million-ton asteroid may be diminished to 2.2 million tons by the end of its journey. Meanwhile people will have prepared a processing plant in orbit to handle the material on its arrival.

Compared with O'Neill's proposals for mining the Moon, fetching asteroids will offer several practical advantages. For a start, the gravity of an asteroid is nearly zero, so there will be no need to 'soft-land' anything, as in the case of the Moon. And while operators relying on the energy of sunlight on the Moon will find themselves unable to work for half the time, during the two-week lunar night, the equipment on the asteroid can remain continuously in sunlight. The quality of the material, especially in respect of water and carbon, will make it more useful than the lunar rock.

O'Leary estimated and compared the costs of putting a ton of material into a high orbit about the Earth: lifted from inside the deep gravity well of the Earth itself, it will cost about $250,000 per ton; quarried and launched from the Moon, about $1000 or $2000; retrieved from a suitable Earth-approaching asteroid, the cost per ton may be as low as $100 to $600. But critics said that asteroid-harvesting was too complicated to leave to robots: if men have to go, the costs will presumably be a good deal higher.

In O'Leary's thinking, the harvest of asteroids will serve in the first instance for manufacturing in space and for building and supplying space settlements. As a supporter of O'Neill's scheme for satellite solar power stations to provide energy to the Earth, O'Leary offered asteroidal rather than lunar material for the purpose. O'Neill himself was the first to propose using a mass driver as a tugboat to bring asteroids to the vicinity of the Earth, but he saw it as a later development, to be pursued only after construction in space, using materials from the Moon, was well advanced. O'Leary, on the other hand, advocated going straight for the asteroids, perhaps dispensing altogether with the lunar mining operation. He went so far as to suggest that the material could be used to build farms in space for the supply of food to Earth, declaring roundly:

'We're really talking about nothing less than a means by which we can retrieve resources from shallow gravity wells. Then the limits to growth in the biosphere of the Earth are expanded and worldwide food shortages, resource shortages and energy shortages are essentially removed, just by the mining of a few fairly small asteroids.'

Be that as it may, the asteroids will offer a valuable source of metals for use on the Earth itself. Michael Gaffey and Thomas McCord at the University of Hawaii, who studied the composition of asteroids from afar by their absorption of characteristic wavelengths of sunlight, came up with a scheme for mining asteroids that was somewhat different from O'Leary's. In particular they envisaged dropping huge pieces of asteroidal metal into the Earth's ocean. Unlike Shoemaker and O'Leary they favoured the main belt of asteroids beyond Mars as the most promising source of useful asteroids.

In the Gaffey-McCord version, an asteroid will be selected for its high content of free metal, and miners will live on it while it follows its natural orbit about the Sun. Rather than shifting the asteroid as a whole, they will mine and refine the free metals and send ingots by tug to the space factories orbiting the Earth. There another workforce will fashion the metals into wing-like 'lifting bodies' capable of half-flying, half-falling through the Earth's atmosphere without melting. Taking advantage of the zero gravity and vacuum in space it will be a simple matter to inject a gas into the metal to produce a metal foam less dense than water. A

Fetching asteroids *to the vicinity of the Earth is the dream of Brian O'Leary, an astronomer working closely with Gerard O'Neill. He was pictured (above) during his training as an astronaut. In O'Leary's opinion it should be more economical to use the material of suitable asteroids for construction in space, than to mine the Moon. Following a proposal of O'Neill, he favours the use of 'mass drivers' to move an asteroid towards the Earth and then into a high orbit around it. A mass driver (model, right) would take rock from the asteroid itself and eject it at high speed, thus pushing the mass driver and the asteroid along, by reaction. A large part of the mass of an asteroid would be sacrificed for this purpose, but the residue arriving near the Earth could still be very great.*

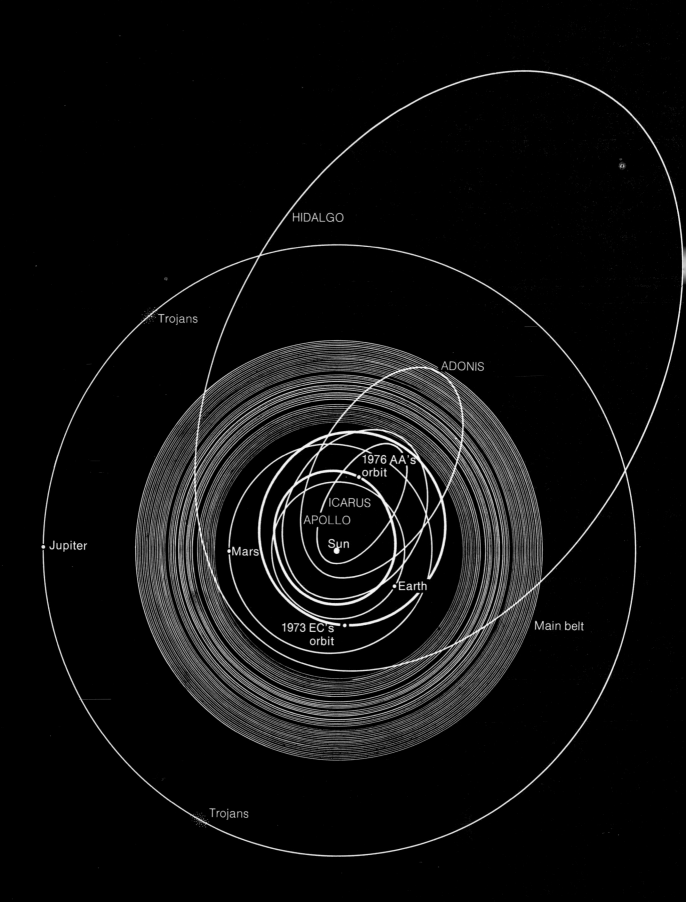

single body for delivery to Earth may contain 100,000 tons of metal. Space tugs will put the body into a trajectory that grazes the Earth's atmosphere, and on its way down computers on board will shift weights about inside it, to correct the flight-path. Finally the metal will come down with a 100,000-ton splash into the ocean. There it will float until seagoing tugs retrieve it, taking over where the space tugs left off.

A cubic kilometre of asteroidal metal, which might be recovered from a single asteroid, will provide 6900 million tons of iron, 800 million tons of nickel, 40 million tons of cobalt and 8 million tons of copper. In simplistic terms, at mid-1970s prices these will be worth nearly $5 million million. But the supply of nickel alone will be equivalent to about 1000 years' consumption of the metal, so the splashdown of asteroidal foam will have a marked effect on prices and on the patterns of use of these materials. Nickel is resistant to corrosion and will find much wider uses, in stainless steel for example, when supplies from space make it cheaper. Incidentally, mining asteroids for metals will not be a complete novelty. Our ancestors' first introduction to iron probably came from picking up iron meteorites. And the great nickel deposits at Sudbury, Ontario, which supplied half the world's nickel in the 1960s, were remains of a large asteroid that hit the Earth nearly two billion years ago. Indeed it may be sensible to look for other existing deposits like that, before chasing asteroids in space.

Iron will never be scarce on Earth, yet importing asteroidal iron need not be an absurdity. Apart from its valuable contents of nickel, cobalt and other metals, the fact that the iron comes ready-smelted by nature will be no trivial matter. In the 1970s, fifteen per cent of all the coal mined on Earth was going into the smelting of iron ore, and the mines, blast furnaces and slag heaps associated with iron production impaired the natural environment in many places. According to Gaffey and McCord, the energy spent in bringing a ton of free iron from a well-placed asteroid will be less than the energy required to obtain a ton of iron from high-grade iron ore in a terrestrial blast furnace.

Behind their cool analysis of the characteristics of asteroids, excitement quickened the minds of the astronomers, as they began to visualise heroic voyages that will take men or robots on year-long journeys to new El Dorados far away in space. By an estimate of Gerard O'Neill's, if the vast resources of the main asteroid belt are ever hammered out into artificial space settlements, they will create new land for human habitation three thousand times greater than the area provided by the Earth. Looked at another way, the asteroids will also be remote places where people can go to live by choice. 'A mixture of asteroidal minerals,' Eugene Shoemaker remarked, 'wouldn't make a bad soil.'

So far from moving an asteroid into Earth orbit, the settlers may wish to drive it to another part of the Solar System to escape outside interference. According to Freeman Dyson, people will settle on the lonely asteroids to escape from an oppressively uniform world on Earth. That is a thesis to be pursued later, when I return to the question: under whose auspices, official or otherwise, will spacefaring ventures be undertaken? But first we should take further stock of the resources of the Solar System and of the means of reaching them. There is five hundred times as much material in orbit around the Sun, as in the Earth, the Moon and the asteroids put together. It is mainly in the huge outlying planets.

Most asteroids orbit the Sun in the 'main belt' between Mars and Jupiter. The Trojans ride at 'Lagrange points' on Jupiter's orbit. But some (examples shown) cross inside the Earth's orbit. 1973 EC and 1976 AA have been suggested as targets for exploration. (After maps by William Hartman and by John Niehoff.)

Amazing robots explored the Solar System while manned expeditions to the planets were still only a vague dream.

Upper left. *Before its flight to Jupiter in 1973, Pioneer 11 was checked against a mock-up of the nose cone of the launching rocket at Cape Canaveral.*

Lower left. *A NASA artist's impression of one of the Pioneer spacecraft passing near Jupiter. It carried thirteen experimental packages and the dish provided communications with the Earth.*

Upper right. Mariner 10 *inspected the inner planets, Venus and Mercury, in an extraordinary bout of planet-hopping that culminated in 1975 in a swoop to within 200 miles of Mercury.*

Lower right. *A model of a* Voyager *spacecraft. Two of these craft were outward bound in 1977, on a joint voyage of discovery to Saturn, Uranus and Neptune.*

7

Two unmanned spacecraft left Cape Canaveral in August and September 1977 for a great voyage of exploration through the outer Solar System. Budgetary trimming made it less than the Grand Tour that space scientists wanted, to take advantage of a favourable arrangement of the outer planets the like of which will not recur until the twenty-second century. But *Voyagers 1* and *2* will take a close look at Jupiter and Saturn and, if *Voyager 1* fulfils that part of the mission, its sister will make a bid to reach the other two large outer planets, Uranus and Neptune. If all goes well,

Voyager 2 will arrive at Neptune in 1990. By then our knowledge of the economic resources of the empire of the Sun will be much firmer.

That thirteen-year timetable for an unmanned spacecraft, travelling so far afield that eventually radio signals will take more than four hours to pass between the Earth and *Voyager 2*, would have seemed ridiculous in the early days of spaceflight. But a succession of complicated flights by long-endurance American and Russian spacecraft explored all of the nearer planets – Mercury, Venus, Mars and Jupiter – with such a remarkable rate of success that there seemed to be almost nothing that ingenuity could not accomplish. While the astronauts took the limelight in the *Apollo* flights to the Moon, the space engineers were the undisputed heroes of interplanetary exploration, and their small robot craft were extraordinary in an age not lacking in engineering wonders. The American spacecraft were particularly reliable. And as each one made its impeccable rendezvous with its target planet, it sent back better pictures and more information than ground-based astronomers had gleaned in centuries of study.

For this novel kind of planetary research, many engineers, technicians, subcontractors and laboratory scientists were involved in packing miniaturised instruments, control devices and communications systems into vehicles smaller than sports cars. From conception to completion a mission might take a decade – with due allowance for official hesitations about budgets running at tens of millions of dollars. There were failures, of course, yet the engineers showed much skill in nursing sick spacecraft.

Take the case of *Mariner 10*. When it was launched in November 1973 it looked as if it was crippled. Guidance gyros performed erratically and cameras were frozen; those were two snags encountered at the outset. But the controllers at the Jet Propulsion Laboratory in California persuaded *Mariner 10* to do everything expected of it. It took a close look at Venus, and swung on a carefully calculated trajectory around the planet, heading on to Mercury, the innermost member of the Solar System. Then it used Mercury's gravity to guide it into an orbit that would enable it to encounter Mercury twice more. On the third pass, in March 1975, *Mariner 10* skimmed just 200 miles over the surface, near Mercury's north pole. After a 'break' like that on a billiard table a hundred million miles wide, the plans for the *Voyagers* looked quite believable.

The *Voyagers'* research involves eighty-five scientists back on Earth and concerns, in part, the flows and shocks of the diffuse solar wind blowing out from the Sun and encountering the magnetic fields of the planets; also the weather on the great planets themselves. But there is intense interest too in the chemical make-up of the planets, their attendant moons, and Saturn's elegant rings. The flight path of one of the spacecraft was devised to take it particularly close to a special object. In November 1980 *Voyager 1* will pass within 5000 miles of a large moon of Saturn, called Titan. For planetary astronomers and would-be space settlers, it is a fascinating object and provides a fitting start to any sketch of the riches of the Solar System.

Titan is the largest of all the moons in the Solar System, being about 3500 miles in diameter – bigger than the planet Mercury though with much less mass. The Dutch astronomer Christian Huygens first spotted it in 1655, orbiting about Saturn every sixteen days, but not until 1940 did his successors begin to find out anything much about it. Then Gerard Kuiper in the USA found that Titan possessed a considerable atmosphere containing methane gas – known on Earth as 'natural gas'. A combination of telescope observations, solar-system theory and laboratory experiments led to firmer ideas about the object, by the time the *Voyagers* set off for a closer look.

Titan has a pronounced red tinge. It is plausibly due to clouds containing mixtures of organic materials formed by chemical reactions in the atmosphere and resembling the 'soup' from which life began on the young Earth. Indeed the early atmosphere of the Earth may not have been very different from Titan's and a closer analysis of those red clouds and deposits on the surface may shed new light on our own origins. But only the most optimistic imagine there might be actual living things on Titan, given a surface temperature of perhaps —150°C. The discovery of hydrogen in Titan's atmosphere, in 1972, was puzzling because the moon's gravity was plainly too feeble to retain the hydrogen unaided. One explanation is that Saturn's strong gravity helps to hold the gas in a hoop around it, and Titan 'borrows' some of the hydrogen as it travels through it.

The most thought-provoking feature of Titan is the constitution of its main body, as deduced from its low density and from a general theory of planetary chemistry. According to John Lewis of the Massachusetts Institute of Technology, the rocky interior of Titan is plausibly surrounded by a deep layer of water containing large quantities of dissolved ammonia – very like a domestic ammonia solution. On top of that floats a thick crust of water-ice with a lot of methane trapped in it. Water, methane and ammonia – those are the inferred treasures of Saturn's giant moon.

If the *Voyagers* and eventual landers confirm that picture, even roughly, Titan will be an amazing source of raw materials for human and other life in the Solar System. The quickest way to sense the potential is to note that we ourselves, to a first approximation, are made of a mixture of water (H_2O), methane (CH_4) and ammonia (NH_3). If you take 102 pounds of water, thirty-six pounds of methane and six pounds of ammonia you have about ninety-nine per cent of all the atoms in a 150-pound human body, in almost exactly the correct proportions. Of the key materials, methane, with its indispensable contribution of carbon, is likely to be the one in relatively short supply on Titan. If Lewis' estimate proves to be correct, that methane constitutes four per cent of the mass of Titan, there will be about five billion billion (5×10^{18}) tons of 'natural gas' contained in the titanic ice. That much methane, with appropriate admixtures of water, ammonia and a residual four per cent by weight of further elements, will be sufficient in principle to 'reconstitute' incredible numbers of people (10^{20}). That would be a dubious objective.

It is more appropriate to think of life as a whole, in the forests and grasslands, rivers and seas of the Earth. The carbon present in all plants, animals and microbes, in the decaying remains of dead organisms and in the carbon dioxide of the Earth's atmosphere and surface waters, comes to about 5000 billion tons. By that reckoning, the resources of Titan will be sufficient to support life a million times more abundant than the Earth's. So much life will not fit on Titan itself; nor will the feeble light so far from the Sun drive it very effectively. But one can look forward to titanic fertilisation of the Solar System, with the precious materials being ferried from Titan to more congenial settlements in orbits closer to the Sun.

Scrutinised from the Earth the rings of Saturn, lying far inside the orbit of Titan, seem to consist of small particles of ice. Whether the ice might contain methane and ammonia, making them a ready-ground version of the resources of Titan, is unclear. In 1977, estimates of aggregate mass of the rings of Saturn were running at more than three billion billion tons (3.5×10^{18} tons) or about $1/20$ of the mass of the Earth's Moon and $1/40$ of the mass of Titan. But already voices can be heard calling for a ban on all economic exploitation of Saturn's rings, for the sake of preserving their ethereal beauty which is unmatched anywhere in our cosmic neighbourhood.

Long before *Voyager 2* set off to visit Uranus and Neptune, astronomers knew that the two giant planets beyond Saturn contained methane in their atmospheres. The structure and composition of these remote objects remains enigmatic. But there are general grounds for supposing them to be large Titan-like objects wrapped in a thick atmosphere consisting mainly of hydrogen. With masses 600–700 times greater than Titan's they presumably embody large reserves of water, methane and ammonia. An easier, low-gravity target for spacecraft is Neptune's large moon called Triton – by Lewis' theory it is richer in methane than Titan.

The unexpected discovery of Chiron, a 'mini-planet' crossing the orbit of Saturn, was announced in November 1977. Charles Kowal, who found it, supposed that it might be the first of a swarm of asteroids lying in a new belt between Saturn and Uranus. If they exist, they are too far from the Sun to be of much interest to early space settlers. In the long run the new asteroids will perhaps provide useful stepping-stones for visiting or occupying the outer regions of the Solar System.

More than two-thirds of all the material orbiting around the Sun is contained in the largest of the planets, Jupiter. It is 320 times heavier than the Earth. Lying just beyond the main belt of asteroids, Jupiter is also the nearest of the great outer planets, where the material of the Solar System changes dramatically in character and abundance. Jupiter and its attendant moons promise extraordinary possibilities for mankind and may become the most important focus of space operations during the twenty-first century. The two *Voyager* spacecraft launched in 1977 were to make Jupiter their first port of call in 1979, but two earlier spacecraft, *Pioneers 10* and *11*, visited the planet and its moons in 1973 and 1974. The pictures and measurements that they sent back to Earth confirmed and extended the previous knowledge of Jupiter, built up by astronomers over hundreds of years.

Although it is very similar in its composition to the Sun itself, Jupiter is thought to be in essence a huge spinning drop of liquid hydrogen. During the origin of the Solar System, it collected on a small rocky core large quantities of light gases, especially the hydrogen and helium which did not persist in any abundance nearer the Sun. Jupiter's main body, 54,000 miles in diameter, consists of metallic hydrogen, formed when hydrogen atoms are subjected to enormous pressure. Electric currents circulating in this curious molten metal equip Jupiter with a magnetic field a good deal stronger than the Earth's. Around the liquid metal another great thickness (15,000 miles) of ordinary liquid hydrogen constitutes at its top the liquid surface of the planet.

Above that lies a dense, stormy and colourful atmosphere, rich in hydrogen, helium and other materials and possessing thick layers of clouds. In ascending order from the surface, they are water-ice clouds as on the Earth; brown-grey clouds made of compounds of ammonia, sulphur and carbon; white clouds of ammonia; and deep red clouds possibly coloured by phosphorus or organic compounds. Storms on Jupiter persist for a very long time: the famous Great Red Spot is now interpreted as a tropical hurricane that has raged for at least three hundred years.

The jovian weather, unlike the Earth's, is stirred not so much by the Sun as by the planet itself. Jupiter radiates twice as much heat as it receives from the Sun. The heat is left over from the formation of Jupiter. Unlike the Sun at the time of its origin, Jupiter did not attain the necessary central temperature for thermonuclear burning to begin. Nevertheless it was very hot, and the heat continues trickling out after more than four billion years because the transfer of heat between the metallic and non-metallic layers of the planet is very slow.

For space travellers there will·be a notable hazard in approaching or even passing Jupiter. Flying near Jupiter *Pioneer 10* absorbed intense atomic radiation a hundred times greater than the lethal dose for humans. The big planet is wrapped with belts of radiation extending for a million miles around it and consisting of particles – protons and electrons – trapped and accelerated by Jupiter's strong magnetism. They are a scaled-up version of the radiation belts around the Earth found by James Van Allen in 1958, in the first major discovery in satellite research.

Of the thirteen moons of Jupiter, four are especially large and interesting: Io, Europa, Ganymede and Callisto, all named after sexual conquests of the mythical Jupiter. Galileo spotted them with one of the first telescopes in 1610, and the little model of the Solar System which they provided was the most telling evidence imaginable for the Copernican theory. On close inspection, Io and Europa seem like rocky bodies, and Io is spattered with large white and orange deposits of salts, which make it the most gleaming object in the Solar System. Io, Europa and Ganymede all lie within the intense zone of the radiation belts; indeed they serve to mop up most of the radiation – but not enough. The fourth big moon, Callisto, at just over a million miles from Jupiter, escapes the worst of the radiation and therefore attracts special attention from would-be space venturers.

Callisto is bigger than Mercury and is light enough in mass to consist mainly of water. By Lewis' theory, there is a lot of ammonia dissolved in it, but not nearly as much methane as in Titan. It is reasonable to think of Callisto as a huge dirty snowball reeking of ammonia. At a daytime surface temperature of —145 °C, the crust should be firm enough under foot. Callisto may well be valued as a source of water and other light compounds essential for life, but its most significant use may be as a base for' the exploitation of the resources of Jupiter itself.

The first great prize from the jovian system will probably be the helium present in great abundance in Jupiter's own atmosphere. Although helium was created in the Big Bang and became, after hydrogen, by far the commonest material in the universe, it is so rare on Earth that scientists discovered it first in the light of the Sun. It has virtually disappeared from the Earth because of its very low boiling point and its chemical inertness. Twentieth-century physicists and engineers have found all sorts of special uses for helium. For example, helium serves in refrigerators for reaching close to the absolute zero of temperature, in which condition various metals and alloys lose all resistance to electric current. These superconductors promise great benefits in electrical engineering. Moreover, liquid helium has curious properties such as superfluidity that could give rise to new devices. If the airship is ever to make a comeback on Earth, it will need helium to lift it rather than combustible hydrogen. But all such thoughts are jeopardised by the meagreness of the supplies of helium on Earth.

Even more tantalising is the rare, light form of helium. Helium-3 is a potential nuclear fuel. The biggest drawback in the plans for generating energy on Earth by the thermonuclear fusion of the heavy forms of hydrogen (deuterium and tritium) is the copious production of neutrons. These nuclear particles are dangerous to life and will damage a nuclear reactor. The virtue of helium-3 is that it reacts with deuterium to produce helium-4 and protons – charged particles which are much more easily arrested than neutrons. For that reason, helium-3 would be the fusion fuel of choice, in preference to tritium, if it were not virtually non-existent on Earth. But Jupiter has plenty of it.

Helium-3 from Jupiter will revolutionise the economics of power production on Earth. That idea is due to Anthony Martin of the British Interplanetary Society. In his opinion, the cost of fetching the helium-3 from Jupiter will not be high in relation to its energy value. One ton of helium-3 is equivalent to about ten million tons of oil. And the cost will in any case be more than

---

Mars, *bleak as it is, is nevertheless the least uncongenial of the other planets from the viewpoint of human settlement. In this picture sent by* Viking 2 *as it approached Mars, huge volcanoes stood out as great pimples, while the white patch at the bottom of the crescent was a sign of frost. Compared with the Earth, Mars is 1½ times farther from the Sun, solar intensity is 43 per cent and the surface area is 28 per cent.*

offset by the benefit of having a nuclear fuel that is 'cleaner' than any available on Earth. The neutron problem will not be quite eliminated, because side-reactions continue to produce some neutrons even with helium-3 as a fuel, but it will be greatly eased. And that will be an important factor, too, when people come to build nuclear-fusion rockets.

The British Interplanetary Society had a special reason for figuring out how to obtain helium-3 from Jupiter. A group in the Society was conceiving a huge spaceship that might make a fifty-year journey to another star (see Chapter 10). It needed helium-3 as a fuel, but ten times more than the total amount of helium-3 on Earth. So a member of the group, Robert Parkinson, came up with a scheme for the mining of helium from the atmosphere of Jupiter.

Robot vehicles rather than manned craft will go into Jupiter, because of the hazards of radiation and gravity and atmospheric pressures much stronger than the Earth's. Any humans involved in the operation may well stay at Callisto. The craft sent to Jupiter will deploy 'hot-air' balloons to float in the jovian atmosphere. The balloons and their equipment will be powered by nuclear reactors and liquefying plant will extract deuterium and helium from the atmosphere. The common helium-4 and the rarer helium-3 can then be separated, perhaps by using the superfluidity of helium-4. On-board radio transmitters will enable the human managers on Callisto to keep track of their movements. Every few months another vehicle will visit each balloon, collect its output of precious isotopes and carry it back to Callisto.

Quite apart from fuelling starships, the helium-3 will be much in demand for power production in fusion reactors on Earth and elsewhere. As another of the interplanetarists, Alan Bond, told us:

'I don't think there's any single resource in the Solar System which will be as important as the helium-3 in the jovian atmosphere. This supply of fuel back to Earth and to other communities within the Solar System will show that the human race is truly opening up this new environment.'

What about the resources of the inner Solar System? As far as supplies to the Earth are concerned, Mercury, Venus and Mars have nothing obvious to offer that will not be more easily obtained on Earth, or from the Moon and the asteroids. Concerning Mercury, the closest to the Sun, planetary theorists gained plenty of fascinating new knowledge from the visit of *Mariner 10*, but there was little to tempt interplanetary prospectors. They already had reason to suppose, from the high density of

Mercury and its proximity to the Sun, that it would consist largely of iron. The *Mariner* pictures showed the surface to be as pock-marked with meteorite craters as the Moon, and a thick layer of rock hid the iron. A civilisation more advanced than ours may one day wish to mine Mercury for its iron, but their machines will have to cope with a dry, airless world where the thermometer climbs past 400°C in the day and drops below −150°C at night.

Venus, the planet most like the Earth in size, is scarcely more attractive than Mercury. Thick clouds of sulphuric acid float in an oppressively dense atmosphere of carbon dioxide. The Russians scored their greatest interplanetary success in 1975 by landing *Veneras 9* and *10* on the planet's surface. These spacecraft, which looked like frogs in panama hats, reported the temperature to be 500°C. Daylight penetrated the gloomy clouds sufficiently for the spacecraft to take and transmit pictures of the stony venusian landscapes from two very different spots 1200 miles apart.

Carl Sagan interpreted the hot, thick atmosphere of Venus as the product of a 'runaway greenhouse effect' whereby the warming effect of carbon dioxide raised the temperature too high for the planet to be able to re-absorb the carbon dioxide in any way on the planet. When the Sun grows old and overheated, the same process will kill life on Earth. But Sagan suggested a far-out cure for present-day Venus: that one might seed its atmosphere with very hardy algae, microscopic plants which could contrive to live in the venusian atmosphere, taking advantage of the traces of water vapour detected among all the carbon dioxide. While they drifted slowly down to extinction in the oven-like conditions near the surface, the algae could conceivably multiply, absorbing carbon dioxide as they grew. When they came to be cooked, they would carbonise, releasing oxygen gas and carbon dust. Thus the algae might, over a long period, reduce the carbon dioxide to its elements. Biologists of the gaian persuasion doubted whether life could operate like that, in penny-packets.

Meanwhile the bright planet Venus, for long pictured as a delightful abode of life, is a greater disappointment for potential settlers than Mars. At least people will be able to walk on the cold desert surface of Mars, without being crushed or roasted. Although the red planet may possess interesting ores in the vicinity of the great volcanoes and canyons, its chief offering is a planetary platform, with moderate gravity and temperatures no worse than Antarctica in midwinter. Will it be made habitable?

One ploy may be to construct huge orbiting mirrors to

Mighty Jupiter, *more massive than all the other planets put together, lies five times farther from the Sun than the Earth does, and rotates every ten hours. It consists mainly of hydrogen and helium, like the Sun itself. The picture above was transmitted by* Pioneer 10, *when the spacecraft was 1,600,000 miles from Jupiter. The Great Red Spot, which is a long-lived hurricane, is visible on the shadowy edge at the left. Conspicuous right of centre is one of Jupiter's four large moons, Io, passing between the spacecraft and the planet. Another Moon, Callisto, lies outside the belts of deadly atomic radiation surrounding the planet, and could become the base for mining helium from Jupiter. According to a scheme of the British Interplanetary Society, nuclear-powered hot-air balloons (model,* right) *would float in the atmosphere of Jupiter, extracting helium from it. Unmanned rocket craft would visit such balloons to collect the product and bring it clear of the radiation belts. Helium is very scarce on Earth, but has many applications, notably for engineering at very low temperatures. Potentially most valuable, though, is the rare form helium-3, which could provide a superior fuel for nuclear fusion, in power generation on Earth or in rocket propulsion.*

Mercury

Venus

Earth

Moon

Mars

Jupiter

Callisto

Titan

Saturn

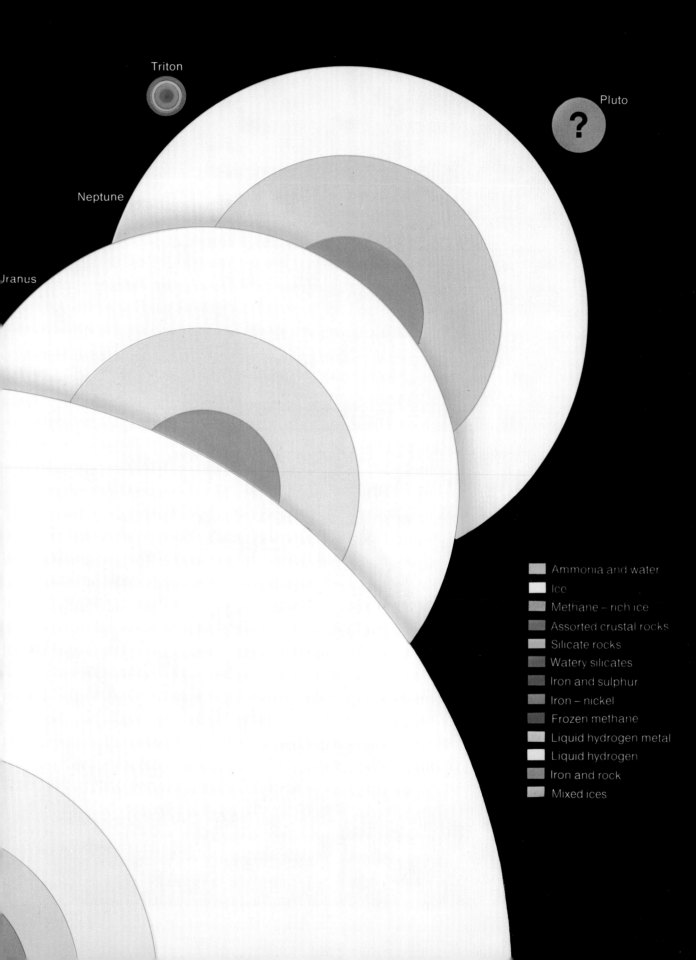

Triton

Pluto

?

Neptune

Uranus

Ammonia and water
Ice
Methane – rich ice
Assorted crustal rocks
Silicate rocks
Watery silicates
Iron and sulphur
Iron – nickel
Frozen methane
Liquid hydrogen metal
Liquid hydrogen
Iron and rock
Mixed ices

focus more sunshine on to Mars, and especially on to its polar icecaps. The water locked up in the icecaps and in frozen soil may prove sufficient, if melted, to cover the whole planet to an average depth of seventy feet or more. To create from the water an oxygen atmosphere as rich as the Earth's would consume about seven feet of that water and a minimum expenditure of energy equivalent to the entire solar power falling on Mars for more than a century. In principle living organisms could do the work there, too.

Heroic deeds of planetary engineering need not involve an undue outlay of human heroism, or even stupendous efforts by microbes. According to Freeman Dyson, a great deal of the work in uncomfortable environments in the Solar System, including preparations for human settlement, will be done by self-reproducing robots. 'I believe this is a big part of our future,' he declared.

Living creatures long ago worked out a system for reproducing themselves. Once human beings understand in detail how the trick is done, they will, by Dyson's reasoning, apply the same principles of organisation in man-made machines built of metal and glass, or possibly even living parts. The machines will have the ability then to manufacture their own parts according to the blueprints and produce complete copies of themselves. They will do anything else that we order them to do as well, when something like Ted Taylor's Santa Claus Machine (see Chapter 1) starts breeding.

If the Santa Claus Machine is to be a very efficient digestive system for consuming rocks and other organic 'food', the self-reproducing version will resemble a more complete living organism. The mathematician John von Neumann was the first to sketch the principles underlying the design of any self-reproducing machine. In doing so he anticipated some of the crucial discoveries of molecular genetics about the way living things pass on operating instructions to their offspring. By engineering based on those same principles of genetic programmes and executive machinery we shall, in Dyson's view, be able to reduce to an egg the essential information for an industrial complex to grow of its own accord. Feeding on sunlight and the rocks of the Solar System, it will be able not only to grow but to repair and renew its various parts. And, when it has grown up, the machine will lay new eggs, which can be taken to hatch at any chosen site.

For colonising space it will be advantageous to send out an egg machine composed of metal and rock which can 'live' in the harsh conditions of space – on the Moon, say, or an asteroid or on Mars. There it will reproduce and, with its 'descendants', will create the basic economic sub-structure for the life of human beings. Dyson foresaw the self-reproducing machines playing their most important part in the human settlement of the inner parts of the Solar System.

Self-reproducing machinery will be of arbitrary scale, to serve almost any practical purpose from chicken-farming to reconstructing the Solar System – people will just write the appropriate programme. So generously will the machines provide for our needs that humans will be able to afford to employ these obedient slaves sparingly and inefficiently, with due regard to aesthetic and ecological limits. For example, they will have to be used with great circumspection on the Earth – perhaps confined to deserts and oceans, where they will not disturb people too much.

Even in space, I think, self-reproducing machines had better be kept under control. If they were allowed to run wild on the Moon they and their progeny could, like so many mice at the cheese, consume our handsome satellite entirely in less than a hundred years. Some people will no doubt resent them or fear them and will always insist on doing work for themselves, with their own simple tools. Nevertheless, the advent of self-reproducing machines will make the grandiose work of exploiting the resources of the Solar System seem easier, or less ridiculous. Our descendants may be glad enough to leave to 'intelligent' and/or self-reproducing machinery the arduous work involved in constructing giant space settlements, in making Mars and the asteroids habitable, or in fetching that helium from Jupiter and methane from Titan.

Before turning back to the twentieth century and the available and foreseeable means of starting all these enterprises, we should not overlook the Sun itself, a far greater source of energy and raw materials than the planets. We cannot rule out the possibility that our descendants will operate very close to the Sun, and even interfere with it in ways we scarcely guess at.

Titan base *was fictional (model, upper photograph) in the BBC television serial* Dr Who. *But astronomers suspect that Titan, Saturn's largest moon, is very rich in water, methane and ammonia. Those materials could support abundant life if they were mined and transpor-* *ted to warmer parts of the Solar System. Saturn's rings (telescope photograph, below) make it the most elegant of the planets. The rings may be a mass of orbiting snowballs containing as much water as all of the Earth's oceans.*

8

CAVORTING IN SPACE

Picture a great installation on Earth with a laser of enormous cost and enormous power, capable of directing a thousand megawatts of light energy upon a spacecraft flying in its beam. It will be a quick-fire launcher, dispatching into orbit a succession of payloads each of about a ton. One day's work will exceed the payload of more than five Saturn V rockets of the kind that send the *Apollo* spacecraft on their way to the Moon. The vehicles boosted by the laser will be smaller by far than the Saturns, because they will not have to carry

Venus, *near-twin of the Earth in size, is a vastly different place. The atmosphere portrayed strikingly in ultraviolet light by* Mariner 10 *(lower illustration) consists of hot and dense carbon dioxide, and the pretty clouds are sulphuric acid. The surface* (upper left) *is littered with stones about two feet wide, in a picture transmitted by the Soviet lander* Venera 9 *(upper right). Radar observations from the Earth suggest that Venus may possess extremely large volcanoes like those discovered on Mars.*

Liquid-fuelled chemical rockets *were the means of breaking the chains of gravity, and they seem set to remain the chief means of lifting loads through the Earth's atmosphere. The world's first liquid-fuelled rocket is shown* (left) *with its inventor, Robert Goddard of Clark University. Powered by gasoline and liquid oxygen, his little craft flew from a field in Massachusetts on 16 March 1926. It attained a height of 200 feet. That was less than the height on the ground of the 2600-ton Saturn V launching rocket* (right) *of forty years later, which sent the Apollo moonships into space. The great bulk was a sign of chemistry's limitations. But once chemical rockets have lifted craft clear of the Earth's atmosphere other means of propulsion in space become possible. Particularly attractive is the idea of solar sailing, and the illustration opposite shows a Jet Propulsion Laboratory scheme for a sailing vehicle which would unfurl its giant blades after launching from the Earth.*

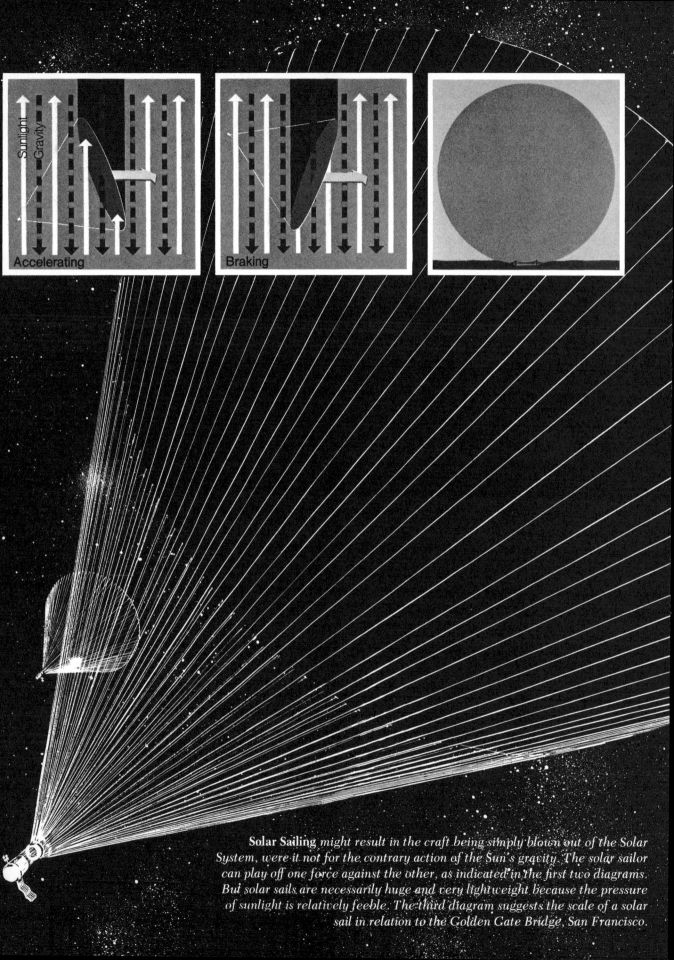

Accelerating

Braking

Sunlight

Gravity

Solar Sailing *might result in the craft being simply blown out of the Solar System, were it not for the contrary action of the Sun's gravity. The solar sailor can play off one force against the other, as indicated in the first two diagrams. But solar sails are necessarily huge and very lightweight because the pressure of sunlight is relatively feeble. The third diagram suggests the scale of a solar sail in relation to the Golden Gate Bridge, San Francisco.*

their own supply of energy – only a relatively small quantity of propellant, less in mass than the payload. The propellant may be water and the laser beam, striking it and heating it, will convert it into a high-speed jet. The acceleration will be about ten times gravity, greater than flesh and blood could stand, but it will be possible to dispatch great loads of machinery and materials from Earth to space by this means, with far less expenditure of energy than chemical rockets would need.

Such was the dream of Arthur Kantrowitz of the Avco Everett Research Laboratory near Boston, a leading laser engineer and another prophet of mankind's break-out into space. Conventional chemical rockets seemed to him an almost absurdly extravagant way of going about it. He liked to point out that the energy necessary in principle to put a pound of material into orbit was only about 4.5 kilowatt-hours – energy which you could buy on Earth for a few cents. Yet, using a chemical rocket of the 1970s, it cost about $1000 to launch a pound of material into orbit. One reason was that the launcher had to lift its own fuel and engines, which weighed far more than its payload; another reason was that the velocity of the jet produced by chemical reactions was not very great. Hence Kantrowitz's wish to put all the effort on the ground and use the laser to export energy to the spacecraft. It could, he thought, reduce costs of launching to one-tenth of the costs of existing rocketry. In a little demonstration in his laboratory, a hollow container attached to a wire revolved at high speed when a laser beam fell on its open end and converted the air inside it into a jet.

This chapter explores some of the ideas about cheaper ways of getting from the Earth into space, and better or faster ways of cavorting about the Solar System. It is not my intention to make an exhaustive review of all the technical concepts for space travel but rather to mention a few interesting possibilities. Their ultimate success may affect the feasibility, cost and organisation of projects for exploiting natural resources in space.

It was upon the Space Shuttle, the advanced American chemical launching system for the 1980s, that Gerard O'Neill and his collaborators relied for their schemes for constructing large space settlements and satellite solar power stations (see Chapter 2). The Space Shuttle's engines represented a substantial and expensive advance in efficiency compared with the Saturn V rocket, and worked close to the limits of available engineering materials. So there was little reason to expect an early improvement on the shuttle technology. For the rent of a Space Shuttle, carrying a payload of twenty-nine tons into low orbit about the Earth, the National Aeronautics and Space Administration proposed to charge $20 million, or about $680 per pound. But there was widespread suspicion that it was not an economic charge and was based on over-optimistic assumptions about recovering and re-using not only the manned orbiter but the boosters as well.

In the belief that the costs of chemical rockets were unnecessarily high because they were 'over-engineered' by government space agencies, a young German rocket expert, Lutz Kayser, and his OTRAG company, set about privately developing a cut-price 'space truck'. It was to be a launcher capable of lifting two tons to a high, geostationary orbit but to be built as cheaply as possible. Kayser planned to make it by clustering hundreds of small rocket engines, each built of ordinary industrial materials and parts and powered by diesel fuel and nitric acid. Despite these unconventional methods the complete launcher, due for its first flight in 1981, was expected to cost hundreds of millions of dollars to develop, and the price tag per launch would still be about two-thirds of the NASA charge for a shuttle flight.

O'Neill, always inclining to optimism, looked to NASA's ideas for developing a space freighter or 'heavy-lift vehicle' using shuttle technology. He thought material could be delivered into orbit at $110 per pound. (Enormous distortions are, of course, involved in any such costings, depending on how much of their expenditure on developing the launchers the governments will write off.) Conceivably though, the hundred-dollar-per-pound chemical launcher will eventually appear. It will then roughly match Kantrowitz's guess at the cost of launching by laser. As chemistry and laser beam are the only methods in sight for getting spacecraft off the Earth, space operations in the foreseeable future will be governed by initial launching costs of at least that order.

Once people or unmanned equipment are in the vacuum of space, quite different and more economical methods of propelling spacecraft will become practicable. In particular they will be able to use the continuous energy of sunlight and the intense energy of man-made nuclear reactions. These may always be denied as energy sources for propulsion within the Earth's atmosphere: solar energy because of the huge, unwieldy collectors needed to absorb adequate power; nuclear energy because of the risks of radioactive contamination of the atmosphere, especially with engines of the explosive type likely to be favoured. In space these impediments will largely disappear.

The electromagnetic gun, or 'mass driver', to be operated by electric power from panels of solar cells, first cropped up in Gerard O'Neill's thinking as a way of shooting soil off the Moon for construction work in space. But then he began to regard the mass driver also as a self-sufficient means of propelling spacecraft, as originally envisaged by the British writer Arthur C. Clarke.

As O'Neill developed the idea with Henry Kolm of the Massachusetts Institute of Technology, the prospects looked bright. Kolm remarked to us about the mass-driver engine:

'It is to the space traveller what the mule was to the early prospectors. It feeds on what it can find – debris and sunlight. You couldn't ask for more from an engine.'

The lunar mass driver is supposed to send buckets equipped with superconducting coils down a magnetic wave and let them shoot their contents into space (see Chapter 3). That concept will remain unchanged in the propulsion mode. The electrical power supply will come from solar-energy collectors. As sunlight is never ending, the mass-driver rocket will remain effective for as long as you have a supply of material that you can throw out of the back of your engine. In other words the spacecraft will still need a propellant – not fuel but 'reaction mass' wherewith to push the craft.

I have already mentioned the use of such mass drivers for shifting the collected lunar soil from one orbit to another, and even for capturing an asteroid. In both of these cases, the reaction mass to be shot from the buckets will come from part of the material being transported. But another source will be spent rocket casings and other 'space junk'. In particular there is the scheme for a so-called 'shuttle upper stage', whereby the empty tanks of the Space Shuttle orbiter will be ground up and fired from a mass driver, to carry payloads from orbits close to the Earth into higher orbits.

The faster the reaction mass is travelling when it leaves the rocket, the more thrust it gives. One alternative to the mass driver, as a solar-powered electric engine, will be the ion engine. It converts its reaction mass into electrified atoms, or ions, and electric fields then accelerate them to extremely high speeds, in a jet. The result is a high impulse from every ounce of reaction mass fed into the electric jet – ten or twenty times more than chemical rockets can give. Because of this efficiency you can use an expensive heavy metal like mercury or caesium as the source of ions. Small ion engines running on solar energy were tested in space in the American satellites *SERTs 1* and *2* from 1971 on-

wards. Later, NASA was planning to cluster a number of mercury-ion engines for a mission to rendezvous with a comet in 1987. The spacecraft may draw its power from huge panels of solar cells.

The next trick will be to use sunlight itself as the reaction mass, in solar sailing. The particles of light, the photons, will bounce off a large mirror, or sail, and thereby transfer momentum to it. The need for a propellant of any kind disappears. While solar sails were being devised that might fly in the 1980s, one enthusiast for solar sailing, Louis Friedman of the Jet Propulsion Laboratory in California, claimed that it was the only propulsion system in sight that could take a manned expedition to Mars and back, before the end of the twentieth century.

The pressure of sunlight is decidedly feeble so the sails for solar sailing will have to be very large and very light in mass. Even then it will necessarily be a very gradual method of accelerating spacecraft. But the unceasing pressure of sunlight in space makes it possible, with patience, to achieve very high speeds. At the distance of the Earth, the force exerted by the Sun on a sail one mile square will be twenty tons and the problem will be to construct sails that are not far too massive compared with the force they gather. The 'solar wind' of high-energy particles streaming from the Sun will *not* contribute significantly to the motive force.

Just as the deep keel of a yacht offers resistance to sideways motion through the water, preventing the boat from simply sliding down-wind, so the 'sideways' motion of a solar sailcraft will be checked by the Sun's gravity. The craft will typically be in orbit around the Sun. Setting the sails to gather energy, one will be able to edge the spacecraft farther out from the Sun; backing the sails to act as a brake will cause it to spiral inwards towards the Sun. The solar sailcraft will thus be able to 'tack' all over the Solar System.

In principle a sheet of metal foil or metallised plastic will make a sail. But because it has to be as flimsy as you dare make it, the sail will be almost impossible to fold without damage. For this reason sails to be made on Earth and carried into space by rocket may be more like a windmill or a set of helicopter blades – the so-called Heliogyro. Once it has been fired on its way to escape the Earth's gravity, such a sail will automatically unfurl its blades from central spools and the whole contraption will slowly spin to provide centrifugal force to hold the blades outstretched. Like helicopter blades, the blades will be pitched at chosen angles to control the spacecraft.

A sail of this type envisaged by JPL will consist, as

Ion engines *in action near Jupiter, as foreseen by an optimistic NASA artist. These unconventional rockets use electric power to eject ionised metal atoms at very high speeds, thus obtaining a high thrust from a small mass of propellant. In the visualised version there are ten electric jets, and the power comes from huge arrays of solar cells. Comparisons with solar sails for 1980s' missions gave the advantage to the ion engines.*

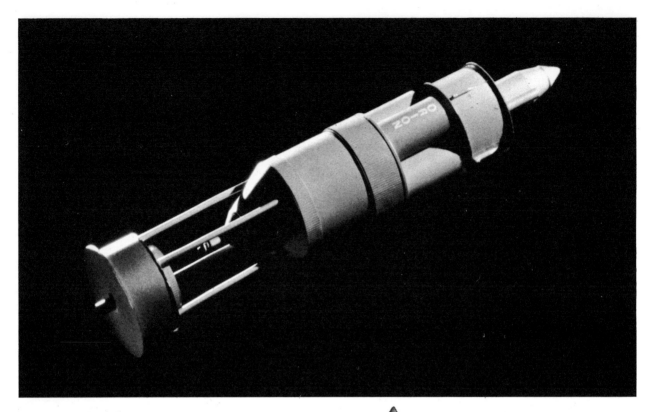

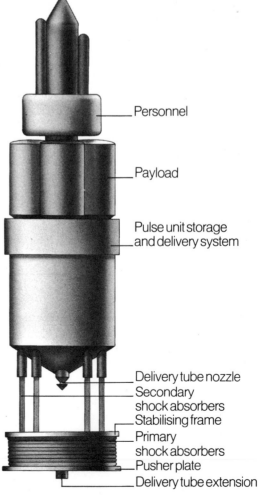

Personnel

Payload

Pulse unit storage
and delivery system

Delivery tube nozzle
Secondary
shock absorbers
Stabilising frame
Primary
shock absorbers
Pusher plate
Delivery tube extension

Project Orion *was an American scheme for propelling manned spaceships by exploding atomic bombs behind them. The model (above) and the diagram (right) show a typical configuration that the Orion team envisaged. The pusher plate, mounted on huge shock absorbers, takes up the force and heat of successive explosions. The system could in principle propel a large, comfortable spaceship at high speed. After Orion was halted by the nuclear test-ban treaty, the idea of using atomic (fission) bombs for 'nuclear-pulse propulsion' was superseded by a preference for small fusion explosions.*

Friedman described it, of twelve blades, each $4\frac{1}{2}$ miles long by 26 feet in width, made of plastic film one ten-thousandth of an inch thick, and aluminised on one side. Edge tapes and crosswise battens will take the strain, and 'rip stops' embedded in the sail every few feet will prevent catastrophic tearing by meteorites. According to JPL calculations, during a four-year mission the accumulation of meteorite punctures will diminish the area of the sail by only one per cent.

Friedman visualised the use of solar sails in unmanned ventures, starting in the late 1980s, to Mars, Mercury and the moons of Jupiter and Saturn. Thus a solar sailcraft might set off for Mars, and there send down a lander carrying roving vehicles. The lander would gather a couple of hundred pounds of martian rock and soil for bringing back to the Earth. Such a mission would take about four years and could pave the way for the first human visit to Mars, using a flotilla of large sailcraft.

Sails built on Earth will not be cheap, although not out of line with the costs of conventional rocketry. But making the sails out in space may bring special advantages. An enthusiastic view of it came from Eric Drexler of the Massachusetts Institute of Technology, who told us that he thought space-made sails could be perhaps forty times lighter, because they did not have to be folded or rolled up. One way of making the sails will be to evaporate aluminium on to a bed of wax, and then evaporate the wax. Strips of very thin metal film formed in that way will then be pasted on a mesh of very thin wires to make huge continuous sails, circular or hexagonal, ten miles in diameter and weighing only twenty tons. One such sail will haul a hundred tons or more of material or equipment at comparatively high speeds around the Solar System. In Drexler's reckoning, carrying material around the inner Solar System under sail will come to rank with carrying it by trucks on Earth.

Solar sailing generates special wonder and excitement among those who contemplate it. However the economics will eventually look to the hard-headed administrators of space enterprises, the prettiness of solar sailing will impel people to attempt it, even if only as a grandiose hobby. And if the technical problems of making and managing the exquisitely flimsy sails can be solved, there will almost certainly be a role for solar sailing as a cheap method of hauling heavy loads about the Solar System, or engaging in freelance voyages of exploration.

But human beings are impatient as well as romantic, and they will not content themselves with creeping around the Solar System on journeys taking years, when they can reach their objectives in a matter of days. Chemistry and sunlight are inherently feeble aids to propulsion, compared with nuclear energy. Pound for pound, nuclear fuel provides millions of times more energy than the best chemical fuel, and as the weight of fuel is what chiefly hobbles the chemical rocket, great advantages will in principle come from the nuclear propulsion. One route to nuclear propulsion might be by way of nuclear reactors, which could be used to heat a jet of gas or to generate electricity. Given electricity you could then run a mass driver or an ion engine. But nuclear reactors and electric generators may well remain so heavy as to cancel out much of the benefit of the high-energy fuel. The simplest and most effective way to exploit nuclear energy for space propulsion will probably be to run a rocket on small nuclear bombs.

According to the participants in the ill-starred American Project Orion you could have high-speed travel to the planets, whenever you wanted it, provided you would tolerate the use of nuclear explosions in space. The concept was taken up by the American government the day after the first Russian *Sputnik* went into orbit. For eight years after 1957, a large team of scientists and engineers explored in all seriousness the possibility of using not-so-small bombs to push along the spaceship named *Orion*. They flew a model using chemical explosives to demonstrate the stability of their unusual means of propulsion; a succession of five cartridges, each with two pounds of high explosive, lifted the model craft 200 feet off the ground.

Theodore Taylor and Freeman Dyson were among the protagonists. As Taylor described their objective to us, they visualised a rather grandiose vehicle able to carry twenty or more astronauts, their life-support systems and their instruments on a trip around Mars, say, or the moons of Jupiter. The ship was to carry perhaps a couple of thousand fission bombs ('atomic' rather than 'hydrogen' bombs) in a big compartment. Released at a rate of one every second, each would produce an explosion equivalent to 10,000 tons of TNT going off, a few hundred feet behind the crew. A large metal pusher plate, mounted on big pneumatic shock absorbers, would take up the force and heat of the explosion and drive the vehicle forward at a steady acceleration of one-half Earth gravity. The Orion team imagined even bigger ships in which several hundred people could travel in comparative comfort.

It was no part of the plan, needless to say, to start up this 'nuclear-pulse' engine on Earth. The idea was to assemble the ship in orbit from pieces lifted from Earth by conventional Saturn rockets. A multiple engine, di-

Electron beams *can carry intense energy, as illustrated by the effect* (above) *of a 250,000-megawatt pulse hitting an aluminium plate for one twelve-millionth of a second. Sandia Laboratories in New Mexico, whose equipment this is, want to trigger thermonuclear fusion by electron beams. The idea is adopted* (below) *in the*

British Interplanetary Society's concept of a nuclear-fusion rocket, where each pellet of fusion fuel, dropping into the propulsion chamber, is exploded by electron beams hitting it from all around. The explosions drive the spaceship away, into the page as it were. Nuclear-fusion rockets could make spaceflight far faster and cheaper.

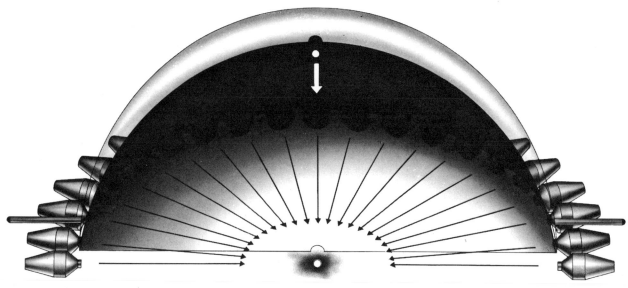

vided into nine units, would in theory carry an 80-ton payload of people and equipment to Mars and back in six months. That can be compared with eleven months for the one-way journey to Mars of the three-ton *Viking* spacecraft launched by chemical rocket in the 1970s.

Orion was cancelled in 1965 after $10 million had been spent – not for technical reasons but as a consequence of the nuclear test-ban treaty. One day someone may blow the dust off the hundreds of secret technical reports that the Orion team prepared. But the concept has been superseded. Any nuclear-pulse rockets of the future are likely to be driven by very small H-bombs. In other words, the thrust will come from thermonuclear fusion of lightweight elements rather than from nuclear fission of heavy elements like uranium or plutonium.

In the years of Project Orion, the only way of starting a fusion reaction was by means of an atomic bomb. That meant the smallest possible release of energy was unmanageably large for anyone except the military people who wanted very destructive H-bombs. But a decade or two later the prospects were brighter for igniting fusion fuel without using a trigger bomb, and so achieving relatively mild explosions suitable for nuclear-pulse propulsion. Alan Bond of the British Interplanetary Society, an engineer experienced both with chemical rockets and with fusion experiments, pointed in particular to the use of lasers or electron beams, for compressing and igniting a small pellet of fusion fuel.

Several laboratories in the USA, USSR and Europe were, by the 1970s, developing laser-beam and electron-beam ignition as part of the general quest for controlled thermonuclear reactions for peaceful purposes. The difficulty, as always, lay in achieving the extremely high temperatures, comparable with those prevailing in the very heart of the Sun, at which the mutual electrical repulsion between atomic nuclei would be overwhelmed, so that fusion could occur. Hitting a pellet of fuel with the beams of laser-light or electrons caused its outer layers to blow off, in the process generating very high pressures on the remainder of the pellet. With ingenious attention to the geometry you could generate a shock wave that heated the pellet to the necessary temperatures for thermonuclear reactions to take place. The remainder of the pellet was then consumed in releasing fusion energy and expanding at a speed of a thousand miles a second or more.

For rocket propulsion, as Bond envisaged it, the fusion pellets will be ignited in a chamber arranged to direct these 'expansion products' in one direction from the spacecraft, thereby pushing the spacecraft in the opposite direction. Many pellets will be exploded in the course of a second, and the spacecraft will carry millions of pellets, thereby achieving extremely high speeds before the fuel is all spent. According to Bond, even if such fusion engines are very inefficient, they will revolutionise spaceflight. Chemical rockets consume something on the order of one pound of propellant for every 450 pounds of thrust whereas in a fusion rocket the same weight of propellant will produce probably 10,000 or 20,000 pounds of thrust. Like the chemical rockets, but unlike the propulsion systems relying on the energy of sunlight, nuclear-fusion rockets will deliver great concentrations of energy from compact engines. So they will reduce journey times around the Solar System in an extraordinary fashion, from the years taken for the *Pioneer* and *Voyager* flights to months and eventually weeks.

Such high speeds will not be mere sops to impatience. The systems needed in the ships to support life for a few days will be far simpler and more economical than those required for journeys taking a year or more. It seems to me that when practical thermonuclear fusion allows us to be spendthrift with energy there will be scarcely any bar to achieving ever-higher speeds. It will be simply a matter of using more fuel and running the engine throughout the flight – perhaps to sustain, for comfort, the same acceleration as that due to gravity at the Earth's surface. For instance the outermost large planet, Neptune, is nearly 3000 million miles away from the Earth. A nuclear-fusion rocket ship, accelerating at one Earth gravity for half the distance and slowing down at the same rate for the remainder of the journey, could make the voyage to Neptune in eleven days. At its fastest, the ship would be travelling at 3000 miles a second. Nuclear-fusion rockets have still to be demonstrated as a practical proposition. Their development will transform the prospects for humanity in space – in principle they open the way to the stars. Nearer at hand, nuclear-fusion rockets will make the entire Solar System *easily* accessible to human beings in space, who have access to supplies of fusion fuel – from Jupiter, for instance, fetched out by robots to the proposed base on the moon Callisto.

9

SPHERES
OF INFLUENCE

Professor: You just sliced the Solar System in two. That one line divides the spheres of influence of the Earth and Callisto. So those little independent asteroid states have been swallowed up.
Researcher: That's correct. The outer and the inner belts of asteroids have both fallen under the influence of Callisto.
Professor: The Earth has still got the planets close to it, under its sphere. But you've got the feeling it's being threatened somewhat by Callisto, haven't you?

Megalithic monuments *asserted local pride and power, according to the archaeologist Colin Renfrew (upper photograph). The great stone structures built in western Europe 6000 to 4000 years ago served as tombs, temples or astronomical observatories. But like the Egyptian pyramids (which began later) the monuments record a habit of engaging in projects that are not strictly necessary in economic terms. Building the observatory and meeting place at Stonehenge in southern England (lower photograph) probably represented, for the time, place and population, an effort as great as the Apollo project for going to the Moon.*

An unusual conversation, you might think, especially taking place between two archaeologists. The professor was Colin Renfrew of Southampton University and the researcher, Eric Level, was running a computer program devised for drawing prehistoric political maps. The rise of Callisto, the moon of Jupiter, in their conjectural model was the consequence of an advance in space propulsion of just the kind envisaged in the previous chapter.

What are the power politics of space colonisation? Under what auspices will the exploration and exploitation of the resources of the Solar System unfold? The eventual answers to these questions will colour the political future of the Earth as well as of the space settlements themselves. We found that the physicists who prophesied the break-out into space had definite but conflicting ideas about how it should be done. Later in this chapter I shall be mentioning the opinions of Theodore Taylor, Gerard O'Neill, Freeman Dyson and others. But we also sought out Renfrew's ideas about the prospects before mankind, because we knew him as an outstanding investigator of mankind's past – and we found him quite willing to speculate about interplanetary relations of the future.

In the late 1960s Renfrew had sent a shockwave through the scholarly world of archaeology by his re-evaluation of the prehistory of western Europe. The barbarians of that part of the world had been civilised, according to the conventional view, by the diffusion of culture and technology from the great early centres at the far end of the Mediterranean. Using newly-corrected radiocarbon dates, Renfrew was able to show that the impressive stone monuments of western Europe, tokens of technological skill and social organisation, were built in fact earlier than their supposed predecessors in the Near East. It was not just a matter of putting the record straight locally: it demolished the archaeologists' framework for describing and explaining prehistory, on the assumption of a gradual evolution and spread of crafts.

Renfrew called for a new approach, in which prehistoric societies were to be interpreted not as benighted communities passively awaiting enlightenment from elsewhere, but as fully functional social systems responding to their local circumstances. He pursued this approach himself in the field, in tracing the rise and fall of civilisations in the Balkans. But he also looked for general principles of social organisation and behaviour that could provide a new framework of ideas for interpreting the usually enigmatic remains dug up by the archaeologist. And if principles enduring

for thousands of years were discernible, might they not be applied to thinking about the future?

One generalisation which Renfrew offered us was that people spent a lot of time doing things that were not strictly necessary from an economic point of view. Most conspicuous for the archaeologist were the great monuments, like the Pyramids in Egypt, Stonehenge in England, and their equivalents in the medieval churches and the great public buildings of modern times. In Renfrew's view, such monuments were assertive statements, declaring in effect 'Kilroy was here'. They often represented, too, competition between neighbouring communities, to see who could put up the better monument. Renfrew thought that Stonehenge was a case in point. The people who built it knew a lot of astronomy and wanted an observatory for looking more closely at the Sun and the Moon. But there was more to it than that: Stonehenge was also a public building, a meeting place where various kinds of exchanges could take place, and an assertion of supremacy. Renfrew commented:

'The effort that went into the building of Stonehenge must in its own terms – in proportion as it were – be every bit as great as the effort that went into the American moonshots . . . And when one speaks . . . of a statement by the Stonehenge people asserting their supremacy, I think there's something of the same again about the moonshots, because what was planted on the Moon was after all an American flag.'

Such enterprises were also symbolic, Renfrew observed, of human attempts to come to terms with the mysteries of the universe. At every stage men made symbols to help them face the unknown. So there would always be cathedrals, space shots or similar imaginative achievements, symbolising a people's unity and aspirations.

His search for generalisations that could guide the archaeologist in the reconstruction of prehistoric social organisation led Renfrew also to the notion of the central place, where people gathered for important meetings. The scale and distribution of central places would be a symptom of a hierarchical political and military organisation. And it was to try to reconstruct social and political boundaries from the patterns of archaeological remains that Renfrew and Eric Level developed the computer program. It took in information about the location and sizes of settlements and then drew the territorial boundaries. For example, it showed how the smaller settlements were probably dominated by larger ones close by, while more remote

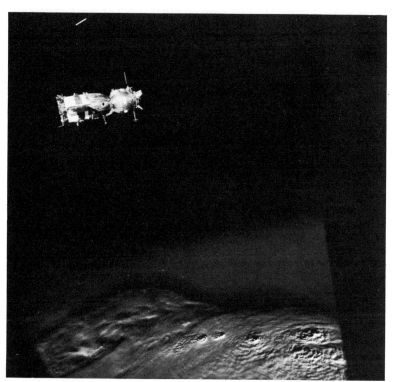

The friendly gesture *of Soviet and American astronauts meeting in space in the* Apollo-Soyuz *rendezvous (1975) may have aided political détente but did nothing to check a continuing arms race in space.*

Left: *The Soviet spacecraft seen through a window of Apollo, high above the Earth. Features of Soyuz include the spherical 'orbital module', the bell-shaped 'descent vehicle' and the cylindrical 'instrument assembly module'.*

Below: *Astronaut Thomas Stafford and Cosmonaut Aleksey Leonov together in the Soyuz capsule.*

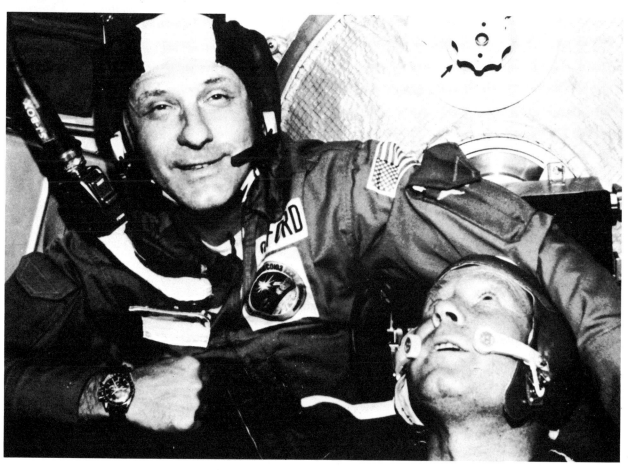

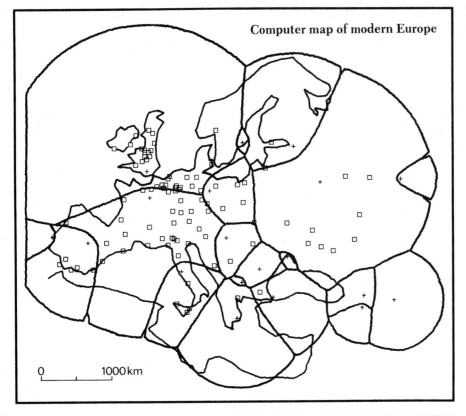

Computer map of modern Europe

0   1000 km

Political power, *present, past and future, is mapped by a computer model developed by archaeologists at the University of Southampton. The model assumes that a settlement asserts political influence proportional to its population but falling away with distance. Present-day Europe* (upper left) *tested the model. The settlements are cities of more than 500,000 people. The computer predicts capitals* (+) *and subordinate cities* (□), *and offers national boundaries. In northwest Europe the model takes no account of the sea barrier when assessing the influence of London, while in southwest and central Europe the large population of Paris has produced boundaries reminiscent of Napoleonic times.*

For prehistoric Rousay (lower left) *off northern Scotland, the population of each settlement is presumed to be proportional to the size of the local burial cairn. The computer predicts independent localities* (+) *and subordinate ones* (□) *within the territory of a larger cairn.*

Hypothetical power-politics in space (right) *unfolds in stages. (1) The Earth* (+) *rules out to Mars, while farther settlements on asteroid Ceres and on Callisto, a jovian moon, are independent. (2) The Earth now controls Ceres and settlers on other near asteroids. Callisto's influence widens because of superior transport. (3) A general improvement in transport makes all settlements subordinate to the Earth or Callisto.*

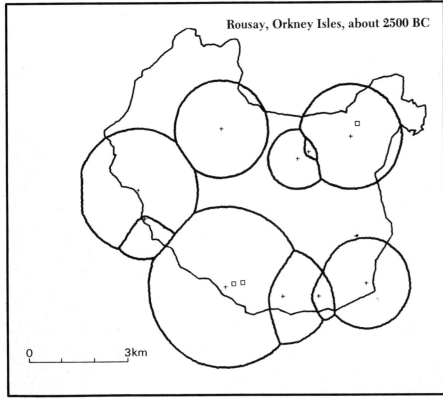

Rousay, Orkney Isles, about 2500 BC

0   3 km

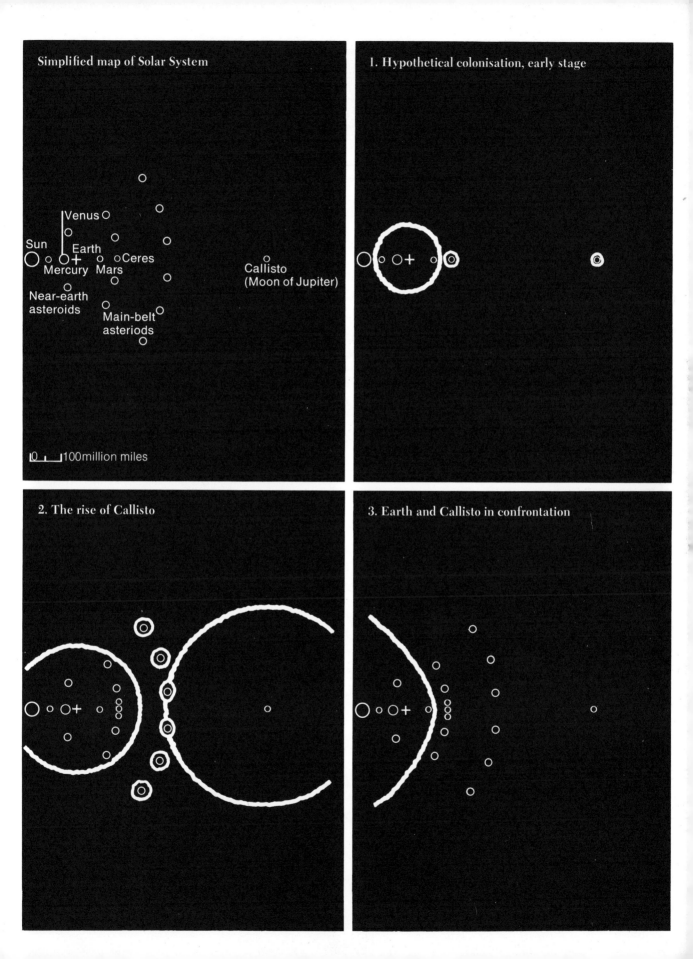

Simplified map of Solar System

Sun
Venus
Mercury
Earth
Mars
Ceres
Near-earth asteroids
Main-belt asteriods
Callisto (Moon of Jupiter)

0 ⎿ 100million miles

1. Hypothetical colonisation, early stage

2. The rise of Callisto

3. Earth and Callisto in confrontation

settlements of the same size could escape such domination. To test their program, the archaeologists fed in data about the 200 largest cities of modern Europe and found that the computer produced a satisfactorily realistic political map of the nations of Europe.

Then Renfrew and Level began applying the procedure to archaeological data, for example from the small island of Rousay in Orkney off the north of Scotland, where there were a great many prehistoric tombs from about 2500 BC. Given the positions and sizes of the tombs, the computer sketched a territorial division for Rousay, such that some territories were served by a single tomb while others had a cluster of tombs. The underlying assumption, of course, was of a certain predictability in human political behaviour. In particular the computer model supposed that great power centres tended to dominate very large areas and that it was usually difficult to set up a little village and hope to escape the influence of the big centres. The people who colonised the Americas were able to escape from the domination of London, Paris or Madrid, because transport was not very easy. In the modern world, Renfrew thought, there was really nowhere to go to escape the influence of the big world capitals – unless people set off to find independence on an asteroid.

But Renfrew also considered it legitimate to apply to the future the regularities found to work for the past. He was quite prepared to speculate that central places in the Solar System will develop their political organisation by much the same rules as those which seemed to have operated throughout history and prehistory. Eric Level used the computer to illustrate the idea. He disregarded in the first instance the curious political effects that will arise from the changes in distances between the planets as they orbit the Sun at different rates, and from the complicated trade-offs in time and energy expended by spacecraft in travelling between settlements. Here is how his completely hypothetical 'history of the future' unfolds.

Stage 1: there are populations on or near to the Earth, Mars, Ceres (the large asteroid) and Callisto (the moon of Jupiter). The computer model predicts, at this stage, a state of affairs in which the highly populated Earth will be fully dominant out to Mars, while Ceres and Callisto are independent states, far beyond the Earth's sphere of influence.

Stage 2: new independent settlements have been established on other asteroids. A technological shift now occurs in favour of the people of Callisto: a new

fuel perhaps gives them greater mobility so that they are able to extend their influence. The computer model represents this advantage by assigning a larger effective population to Callisto. It begins to threaten the independence of the asteroids.

Stage 3: the situation develops with a general improvement in transport efficiency. The asteroids fall under the domination of Callisto, and its sphere of influence meets the Earth's along a single line. The whole Solar System is divided between two superpowers, with the inner region out to Mars dominated by the Earth, and the rest by Callisto, which is growing in power all the time.

Renfrew and Level's cartographic story shows in a schematic way how ancient issues of political domination and rivalry, which antedate Stonehenge, may dog mankind all across the Solar System. Even though the details are entirely conjectural, I fear that the sense of it may be more realistic than some of the hopes held out for an altogether freer and more innocent life in space. Another antidote to naivety comes from the strong whiff, even in the twentieth century, of old antagonisms being extended beyond the Earth.

The *Apollos* were a direct response by the USA to a perceived Soviet challenge in space. And behind the well-publicised civilian exploits of the Russians and Americans a great deal of military work also went on above the Earth's atmosphere. Twenty years after *Sputnik 1* there were about 800 satellites in orbit, of which British radar trackers estimated about fifty to be spy satellites of various kinds. Other spacecraft were playing essential parts in military communications, meteorology, navigation and the guidance of ballistic missiles. By that time too 'hunter-killer' satellites for inspecting and destroying enemy satellites were said to be under test by the Russians. The Pentagon was expected to 'buy' more than a hundred flights of the Space Shuttle, which could also play an anti-satellite role.

People began to visualise full-scale war in space, with the outcome of terrestrial quarrels being decided by who could blind his opponent first by destroying his key satellites. Those might well include civilian satellites, and the extreme vulnerability of satellite solar power stations to attack was one of the strongest objections to Gerard O'Neill's scheme for meeting the world's energy needs. So it was not a question of whether human beings might export their terrestrial rivalries into space some day. They were already at it and the problem was to know how to prevent a large-scale arms race in military

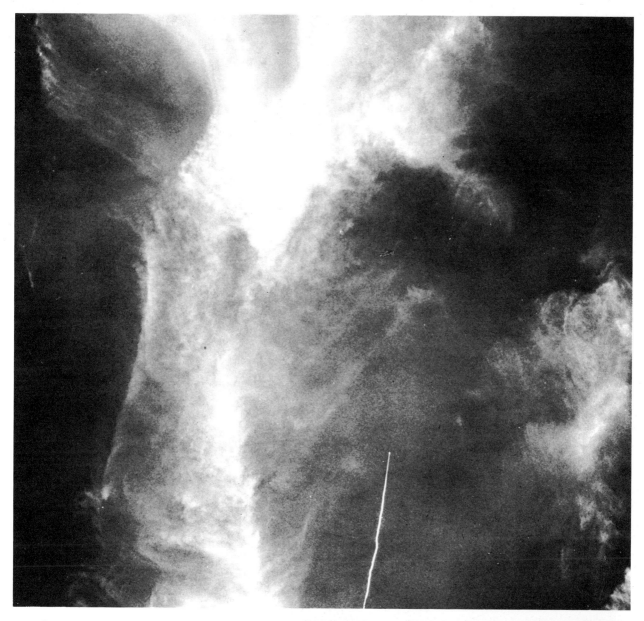

Missile-spotting *potentialities of satellites are symbolised by the picture (above) from the US civilian space programme, where the astronauts in Gemini 7 were able to photograph the trail of a US Navy Polaris missile launched from a submarine off the Florida coast. In practice unmanned spy satellites watch for missiles with instantaneous sensors. Another civilian hint of military possibilities in space appears in the NASA painting (right). It shows a Space Shuttle orbiter fitted with a grab that can recover satellites. The system could serve equally well in kidnapping foreign satellites.*

spacecraft and space weapons accompanying any settlers into space. A cynical but not implausible view would be that the military and political rivalries promise a more effective driving force for the spread of human life beyond the Earth than all the well-intended philanthropic propositions of the physicists.

During his work on the Orion scheme for nuclear-pulse propulsion (see Chapter 8), Theodore Taylor hoped that it could be presented to the world as an international project. That idea arose from a conversation with the great Danish physicist Niels Bohr in 1960. Once he understood the principle of Orion and was persuaded that something of the kind was workable, Bohr began to dream of large-scale exploration of the Solar System in ships bearing the flags and carrying the people of all nations. The world would unite in the adventure of exploring space.

For Taylor, Bohr's dream was not only an appealing vision but an obligatory one for human beings moving away from the Earth. Unless it is to be a friendly, co-operative effort, people will compete for possession of some part of the Moon, some particular asteroid or some position within the rings of Saturn. And then the nightmare of the nuclear holocaust will be extended into space – although Taylor did remark to us parenthetically that, if people had to fight, space was better for war than the Earth was.

That the great adventure of space should involve the whole world, and not be left to the superpowers, was an opinion shared by Alan Bond. He regarded his scheme for nuclear-fusion propulsion for spacecraft as a development of the kind in which his own country, Britain, should invest at once, if it was ever to graduate from regarding spaceflight as a spectator sport. In 1977 very powerful laser-implosion equipment came into operation at the British government's Rutherford Laboratory, and that facility could, Bond thought, serve for a preliminary look at the physics and engineering involved; thereafter Britain could have a definite stake in deep-space exploration, by way of nuclear-fusion propulsion.

At that time Britain seemed to be systematically reducing its commitment to space technology and science, despite earlier successes and the existence of a budding space industry. But other countries like China, Japan and France were offering more serious challenges to the US–Soviet monopoly of space while the West Germans were investing heavily in technology directly relevant to working in space. They were the majority 'shareholder' in the European Space Agency's *Spacelab* – intended to be, after the American *Skylab*, the next practical step in making life out there seem normal, and in developing technological processes for use in space. *Spacelab* was designed to fit snugly into the cargo bay of the American Space Shuttle, accommodating a wide range of experiments and scientists to tend them. The Federal German government was meeting fifty-three per cent of the cost of *Spacelab*, and a Bremen company was the prime contractor.

The European Space Agency was pooling the efforts of West European countries for the development of scientific satellites, applications satellites and the launcher called Ariane, as well as *Spacelab*. It represented a possible arrangement for governmental sponsorship of space activities, intermediate between individual national efforts and the worldwide partnership envisaged by Bohr. But in the West, and especially in the United States, the idea of private enterprise in space was an explicit aim. The possibilities discussed ranged from multinational corporations, through individual national corporations or consortia, to small entrepreneurs who might hope to find in space a new independence and freedom from bureaucratic control.

Consider O'Neill's imaginary picture of the ownership and management of his space settlements and of the satellite solar power stations they will build. In his book *The High Frontier* the imaginary letter-writer in Bernal Alpha explained:

'Legally, all communities are under the jurisdiction of the Energy Satellites Corporation (ENSAT) which was set up back in the 1980s as a multinational profit-making consortium under UN treaties. ENSAT keeps us on a fairly loose rein as long as productivity and profits remain high – I don't think they want another Boston Tea Party.'

Perhaps it will be like that. But wrapped up in this little quotation are some propositions that many governments and citizens will find hard to accept. It implies the creation of a huge commercial monopoly with a 'GNP' comparable with Germany's, and possibly beyond the constitutional control – UN treaties notwithstanding – of the peoples of the Earth. Nations obtaining their electricity from space will not directly control its pro-

The launching rocket *for a Soyuz spacecraft, at Baykonur. The Soviet approach to more elaborate space operations and giving cosmonauts more room in orbit* *involves more than one spacecraft docking at a Salyut space station, while unmanned cargo ships ferry supplies from the Earth.*

duction. In the case of any dispute ENSAT will have the sanction of switching off the supply, much as the Arabs reduced oil shipments to the industrialised world in 1973–74. Finally, the Boston Tea Party analogy suggests that ENSAT will one day lose control of its own workforce and then, perhaps, the power to hold the Earth to ransom will pass to the space settlers. Thus the trouble with proposing schemes for space settlements to bring colossal benefits to mankind may be that the consequential problems and dangers of mismanagement will also become colossal. The price of a reliable solar satellite power system, or any analogous enterprise, might be thoroughgoing world government able to cope with ENSAT and its loose-reined employees and to protect the stations in orbit.

Columbus and his like advertised glorious loot and the glorification of God as dual objectives for their ventures. Similarly running through Gerard O'Neill's studies, and other discussions of settlements in space, were two almost contradictory motives. On the one hand there was hard 'sales talk' about the return on investment for the people on Earth who were going to finance the great enterprise – a return in the form of an energy beam or asteroidal stainless steel splashing into the ocean. In sharp distinction was the softer promise of a new life for the space settlers themselves, a fresh start of the kind that humanity craved down the ages. Unless the US or some other government decides that space settlements are worthwhile for their own sake, these conflicting motives will have to be reconciled and the settlers will have to clear their debts with the Earth before they can enjoy much independence. That will not be new – the Pilgrim Fathers arrived in America heavily in debt to loan sharks in London, and many another immigrant indentured himself for his passage.

Even in that transitional period, conventional accountancy may be hard to apply in space. The settlers will be creating wealth for themselves with their own hands and tools, from solar energy and the materials of the Solar System. When the debts are discharged and the settlers have learned self-sufficiency the umbilical cord from the Earth can in principle be severed. Then, except for such trading as will seem mutually convenient, creation of new settlements can continue at no cost and no return to governments and investors on the Earth. It will even become easier to breed people in

space, than to pay for their transport from the Earth.

That is the theory. In practice, governments and private corporations involved in the initial creation of the space settlements will probably try to remain in charge. They will have a strong whip, in their control of transport to, and supplies from, the Earth. Faced with tough-minded bosses, the space settlers may be less free than at home.

On the other hand it is not hard to imagine individuals seeking to make private fortunes in space. One can paint a romantic scene. Small groups of venturers may go off for years on end, like the circumnavigators who brought spices to Europe in the sixteenth century. They will seek out the vast treasure of an asteroid and settle on it. Then they will manufacture solar sails, from the resources of the asteroid itself, and steer it towards the Earth. Visualise, then, an asteroid weighing millions of tons, rigged with enormous shining sails like a great galleon, voyaging slowly through the ocean of space. The crew dreams of the fortunes to be made at the nickel auctions, when they have trimmed the sheets and glided into their orbital moorings high above the Earth. But will the earthlings tolerate a revival of freebooting capitalism of that kind? And what authority, one wonders, will make quite sure the asteroid sailors do not bungle it, and crash on a terrestrial city?

Perhaps they, or other spacemen, will do it on purpose. In the event of confrontations between people in space and on the Earth the space settlements, whether on the Moon, on asteroids or in orbit, will be easy to destroy. Their pressure vessels keeping out the vacuum of space will be, to nuclear or even high-explosive weapons, as fragile as eggs. But the spacemen will hold the high ground and, like the Cyclops of old, they will be able to hurl rocks at the earthlings. Guided asteroids hitting targets on Earth will have an explosive force of many tons of TNT per ton of rock.

When anyone with a sense of history contemplates colonial expansion into the Solar System, his thoughts keep going back to that European expansion across the oceans of the world, from the fifteenth century onwards. There will of course be no Indians in the Solar System, either to suffer pillage and deceit or to hit back. The European colonies that took over the people and their economies as 'going concerns', in Spanish America, British India and the Dutch East Indies, are

US flags *planted on the Moon by the visiting astronauts told of a national accomplishment. Despite declarations about coming 'in peace for all mankind', Apollo was conceived in fear of the Russians and grew into an* assertion *of American pride and power. On the last of the* Apollo *landings the astronaut-photographer aligned the flag with the distant Earth, as if to reinforce the symbolism of the ceremony.*

not very good analogues for settlements in the deserts of space. The North American colonies that minimised their dealings with the Indian tribes perhaps provide a better model – but even the Pilgrims, when in desperate straits, learned from the Indians how to grow maize, and paid off their debts in London with beaver pelts obtained from the Indians. More realistic in some aspects, as a precursor to the human expansion into space, was the prehistoric peopling of the South Pacific islands. There, as in space, the distances were dauntingly great, the targets small and unoccupied, and the risks high.

In the nearer reaches of space and in the first phase of the break-out one may expect some economic parallels from the European experience, especially while the colonists look back to the Earth for their main trading links. Even if mismanagement from the home capitals prompts political insurrection in the colonies, economic independence may come more slowly. It will be prudent to remember, too, that piracy, war and the unabashed mercantilism of the heavily armed East India Companies were concomitants of the European break-out. The casual hijacking of someone else's sailing asteroid, contests between space navies for the control of special orbits, and the despoliation of moons and planets in the name of commerce: none of that can be ruled out.

I have mentioned more than once the Pilgrim Fathers who left England for America in the seventeenth century. Freeman Dyson, inventor of the Dyson Sphere, offered us a comprehensive thesis about the Pilgrims and their equivalents in the space age. His starting point was a fear of the way the Earth was going. The idea of a united Earth often seemed like a promise of a Golden Age to come, when men would live in brotherhood and tranquillity, settling their differences in the sober chambers of a world government rather than on the battlefields. For many would-be peacemakers, the anxiety was that this utopian dream might never come to pass. To others like Dyson, who were most familiar with the deadly properties of nuclear bombs and the problems of controlling them, the fear was inverted: for staying alive in the nuclear era a united world will be necessary and inevitable – and maybe we shan't like it when we have it.

During the nineteenth and twentieth centuries the tide flowed strongly for large-scale industry and technology, bringing ever-increasing power for the bureaucrats of the state and the large businesses. It curbed the initiative of individuals and imposed uniform ways of life. The great industries spread their standardised products around the world, while the techniques of transport, telecommunications, weather forecasting, public health and environmental protection enmeshed the nations in ever-multiplying international procedures and regulations. World government existed – the only question was, how much?

The nuclear-safe world of the twenty-first century will entail, in Dyson's opinion, the policing of the entire planet, and wholesale destruction of cultures and traditional political freedoms. When originality is too dangerous to allow on this planet, Dyson's argument went, the escape into space will be essential to preserve it. To avoid general suffocation produced by 'bureaucracy, centralisation and paperwork of all kinds', many people like himself will want to go off in search of independence in space, in the manner of the Pilgrim Fathers. But they will be able to do so only when the costs of spaceflight come within the resources of a group of citizens like those who travelled in *Mayflower*.

With this thought in mind, Dyson examined the history of the *Mayflower* venture and came to the conclusion that it was not cheap. The people concerned had to work very hard to raise the money for it and had persistent difficulties with their finances. Dyson estimated that it cost the equivalent of about $20 million, in modern money, for a hundred people with their necessities of life. In other words, it was almost as expensive to cross the Atlantic in the early part of the seventeenth century as to take the same load into orbit in the twentieth century. (By coincidence, the cost of chartering a Space Shuttle for one trip in the early 1980s, carrying a load of 29 tons, is being quoted at just $20 million, at the time of writing.)

Transatlantic voyaging began with national enterprises like those of Isabella of Spain. *Mayflower* and other small enterprises came about a hundred years later. Similarly, Dyson's private colonisers of space will be able to obtain secondhand ships and secondhand experience from the big projects, for carrying out their small enterprises. Given reasonable progress in space technology, the first of them will be able to set off in the twenty-first century.

The asteroids will be the prime objectives, and there will be no special technical problem about reaching them. Survival will then be a matter of biology, in Dyson's opinion. The settlers will have to make do with very little in the way of materials and equipment – far less per head than will go into large settlements of O'Neill's kind. *Mayflower* carried only two tons per person, and the key items in the inventories were seeds, chickens and so on – living things that could reproduce themselves on arrival. 'Cheap' colonisation of space will

'**For sale,** *long-haul freighter, 200 tons cap., only 900,000,000 miles, easy terms.' Pilgrim Fathers of the twenty-first century may set off in a second-hand spaceship (model,* upper illustration*) to seek freedom on a remote asteroid, much as* Mayflower *crossed the Atlantic in the seventeenth century. The relative costs may be comparable, according to Freeman Dyson. At the opposite extreme from the small independent group, other scientists look forward to fully international space projects (model,* lower illustration*).*

Joint European ventures *for the 1980s include the Ariane launcher (model, right) and Spacelab, a versatile laboratory that slots into the US Space Shuttle (below). Spacelab's interior appears in a mock-up (bottom).*

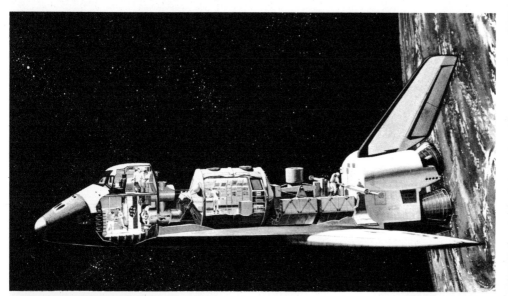

only make sense if a similar mass of material will suffice. And that means finding asteroids with sufficient water, where the settlers can put up a greenhouse and start growing crops right away.

Who will have the courage to go out privately to settle in space? 'All sorts of people,' according to Dyson, 'just as there were all sorts of people coming across the oceans in the seventeenth century.' They will include religious fanatics and people wanting to establish communes of various sorts; also family groups. Dyson once asked a man living with his family on a small island off the coast of Canada whether he would consider colonising an asteroid if he found himself crowded on Earth. 'I don't mind where I go,' he replied, 'so long as I can see what I have done at the end of a year.'

Perhaps I have now mentioned enough possibilities for the reader to develop his own views on how the colonisation of space might or should proceed. But there are other factors outstanding from previous chapters that help to complicate or possibly even clarify the picture. In the absence of any method of leaving the Earth except by large chemical rockets (or possibly the huge laser launcher) the turnstile into space will be under the control of big organisations – international or national administrations, or large corporations. Once people are in space, the going will be easier, with mass drivers, ion engines, solar sails and the like. But the advent of nuclear-fusion rockets will make people on Earth more anxious than ever to keep the spacemen under national or international control. Furthermore, according to the scanty international space law of the 1970s, the parent nation or group of nations will be fully responsible for all space activities originating from them. The independent space settlement may thus be technically illegal, unless the law is changed.

And if the biologists are right in suspecting that small space settlements are not viable ecologically, then small groups have a problem: they will need to become large groups able to create a big settlement, before they cut the umbilical cord. For the immediate future, then, everything seems to be stacked against independence and the fresh start. I suspect that these aims, which appeal to many people, cannot be pursued until large space colonies created for utilitarian reasons have evolved to an advanced stage. Perhaps they will then be producing children who do not have to go through the expensive turnstile, and will be already self-sufficient to a close approximation.

For this reason O'Neill's aim of pushing many people into space as quickly as possible is questionable, on social grounds as well as in matters of feasibility. The settlers will be the servants of governments or profit-making companies. For both kinds of sponsor, independence for the settlers will be something to stop rather than encourage. And the profits will drain back to the investors on Earth rather than going into building ever-more spacious and comfortable colonies. I fear that young people dreaming of a fresh start will find the conditions more like imprisonment on an offshore oil rig or in a polar mining colony, than like the enticing NASA paintings of O'Neill settlements. As I have remarked earlier, it may be better to let a minimum of people and a maximum of robot slaves do the dirty work – that is why I have treated the Santa Claus Machine as a more fundamental concept than the space settlement itself. The day will come when large settlements are easy to build, or to scavenge, as a byproduct of other space activities. Then people can go to them with less obligation to their masters on Earth; then, too, the earthlings had better keep an eye on the Callisto colony.

Paolo Soleri, *the utopian architect, appears* (below) *surrounded by young enthusiasts at Arcosanti, the compact city that he (and they) are slowly building in the desert near Phoenix, Arizona. He aims to exploit to the full the desert's greatest resource – the Sun. Soleri's concept of an orbiting city* (right) *puts a population of 70,000 in a double-skinned inflated cylinder. It is not so much a space project, more a declaration of beliefs about solar energy, living matter and the human spirit.*

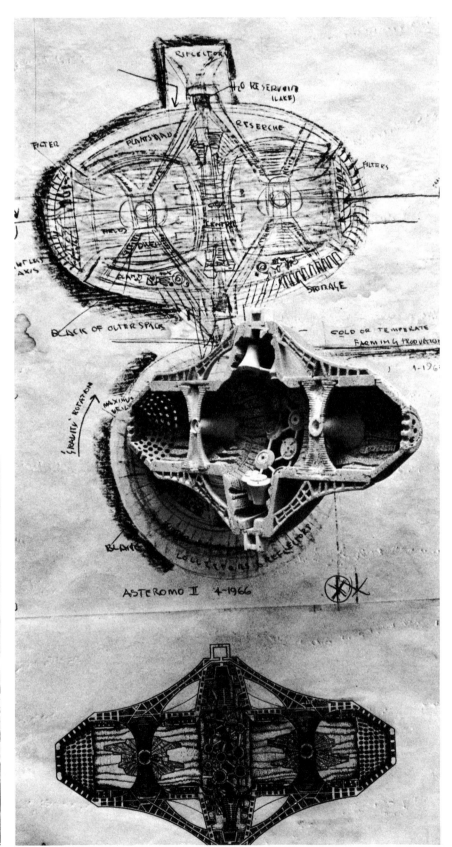

10

THE LOGIC OF DIVERSITY

Since the earliest stone age, even before our own species had fully evolved, human beings were forever trying out new ways of living. Diversity of culture was a dominant theme. Leslie Freeman, palaeolithic archaeologist at Chicago University, spoke of sites he had studied in Spain where, 400,000 years ago, our ancestors came in groups perhaps a hundred strong to drive elephants, horses and other large animals into swampy ground. There they killed them and shared out the meat. But at the same time, he said, smaller groups of people in Syria and China were exploiting different resources in quite different ways, with different kinds of tool kit. Life as a hunter did not set a narrow track for us. It encouraged independence and flexibility, and the ability to meet new challenges. One could see people in the twentieth century adapting from rudimentary hunting or subsistence farming to become clerks and engineers within a generation. Freeman said:

'I believe we're inclined if anything to under-estimate human capacity for adapting to a variety of circumstances – as much as we are to over-estimate that capacity . . . Cultural diversity has been one of the major means of adaptive success for our species and this diversity is now becoming an endangered resource. Encouraging and nurturing cultural diversity is going to be one of the major challenges facing mankind, as we go on to new ways of life we can scarcely imagine today.'

Cultural diversity has meant people leading lives as different as polygamous cannibals and vegetarian monks, Arab millionaires and the members of Chinese communes. Its existence is the chief reason why I believe that social progress is possible: that life for ordinary people can be better in the centuries to come, provided we understand human nature well enough. Learned men have for centuries fumbled this and other issues that ordinary people take in their stride. Society would scarcely function if we were not all shrewd social psychologists, able to predict fairly accurately the behaviour of our fellow humans, in our own cultures at least.

There are paradoxes here. The very diversity of human social behaviour makes it hard to find generalisations about it – for example of the kind being sought by Colin Renfrew in the previous chapter. And now it turns out that one of the key generalisations is the importance of diversity itself. In this chapter I want to consider the logic of diversity from two points of view: utopian architecture and evolutionary biology.

Paolo Soleri became the best-known utopian architect of the 1970s. Born in Italy, he went to work in the United States with Frank Lloyd Wright – the prophet of mile-high buildings. Soleri was sketching giant space settlements a decade before Gerard O'Neill. Under the rubric of 'arcology' – a marriage of architecture and ecology – he produced imaginative designs for cities of the future. One was a combination of city and dam, straddling a canyon with a structure a mile high and accommodating more than a million people. Another was a floating city.

Soleri was a self-confessed worshipper of the Sun. In the 1970s, he and an amateur workforce were building a veritable sun-temple for human habitation in the Arizona desert near Phoenix. Called Arcosanti, it was to be a 25-storey structure covering ten acres of ground, a compact settlement for 3000–5000 people. They were to grow their food in greenhouses and produce metal-work and ceramics for sale. Soleri's design included cave-like 'apses' for admitting the gentle sunshine of morning and evening while keeping out the over-intense rays of the midday summer Sun. Greenhouses would become a source of heat as well as food for the community, with chimneys conveying the warm air to the living spaces;

while evaporation of water provided cooling as required.

Behind such practicalities lay Soleri's theology. Inspired by the mystic evolutionist Teilhard de Chardin, he saw cosmic evolution as a process for transforming matter into spirit, for minimising material needs while heightening the role of consciousness. The city's tightly knit community was the necessary instrument for human evolution.

Human settlements will implode, in Soleri's view, becoming more compact so as to encroach as little as possible upon the wilderness. That will call for ingenuity of the kind attempted at Arcosanti, aimed at reconciling the nature of solar energy, thinly spread over the flat surfaces of the Earth, with the human talent for intensive activity concentrated in three-dimensional structures. 'Crowding' in Soleri's language is virtuous, because of the scope it offers for interaction between people. It will also be 'frugal', again in a positive sense, and will reduce the effort required for transport and communications, as compared with a two-dimensional, sprawling city. Miniaturisation is the goal, with higher levels of complexity to be achieved within a diminished space.

Soleri's creed certainly engaged the imagination of the people who were helping to build Arcosanti, and of the others who admired him from afar. Yet there may be something old-fashioned about his vision of the future. The assumption is that you can conceive an ideal existence for a large number of people and embody it in a monolithic design. The quest for compactness is an old one, since ancient and medieval cities encased themselves in defensive walls. But it took a special form in the architectural thinking of the 1960s. The tag was 'megastructure', meaning a single building containing most or all of the activities of a city: living accommodation, trade, arts, religion, and so on. By the 1970s the idea was somewhat discredited.

According to the architectural historian Reyner Banham, the idea of megastructures was to give a comprehensible architectural shape to the apparently disorderly mess which the car and chaotic land values were making of the centres of cities. The ocean liner was the inspiration for much megastructural thinking, and the idea might well be folded back upon itself in the creation of liner-like cities for ocean habitation. For Banham, megastructures that were actually built tended too often to be simply a stack of housing with some shops slotted in at the bottom; there was little regard for 'undesignable' community spaces.

Banham preferred the more accidental megastructures, like Grand Central Station in New York City, with its complex of railroads and subways on different levels, connected continuously with hotels and office blocks, including the large Pan American building – adding up to a city within a city. He regretted the discounting of megastructures, at the very time when supplies of energy were becoming tighter and a case for compact, energy-conserving cities was developing. Even the long-standing idea of throwing a great plastic blanket right over Manhattan, so that the city could have its own weather inside, might look sensible as time went by. The concept need not be a giant plastic bubble containing buildings, but could be a matter of air-conditioned streets.

Although Banham was looking to a revival of such thinking, it seemed to me that his earlier complaints, in his book *Megastructures*, had identified a central flaw: namely that a city is far too complicated a system, serving far too many practical, visual and cultural functions for any single brain to design it comprehensively. To hear propositions of a completely different kind, we talked with a Japanese-born anthropologist, Magorah Maruyama of the University of Illinois, about the design of space colonies. Maruyama became involved, in 1975, in the National Aeronautics and Space Administration's studies of Gerard O'Neill's proposals for the colonisation of space. In Maruyama's hands the space settlement became, in effect, a blackboard on which to sketch ideas about social systems, human behaviour and the very nature of design itself. And he looked forward to real space settlements as experimental societies where social ideas could be tested out – not just exclusive, single-minded utopian recipes but systems for the cultivation of diversity.

Maruyama brought to the NASA studies an anthropologist's awareness of the differences among human societies. He told us that he found the physicists and engineers more receptive to his ideas than social scientists generally were. One question they discussed was: what *shape* should a space settlement be? There were engineering constraints, such that spheres would be easier and cheaper to build than other shapes. Nevertheless people visualised a variety of shapes besides a hollow sphere, for example a cylinder (as in O'Neill's early conception), a torus (like a hollow doughnut), or a necklace of small living spaces linked together. Maruyama commented on these possibilities from a social-scientific point of view. First, he noted the psychological problem of 'solipsism' – the feeling that everything is a dream and there is nothing real outside one's own mind. Having experienced it himself in

Sweden in winter, Maruyama thought that solipsism might generate apathetic or capricious behaviour and become a serious problem for people in space communities.

Colony designers can seek to avoid such feelings of unreality by having landscapes that look larger than a stage set and in which unpredictable things go on beyond full human manipulation – changes in the weather, for example, or movements of wild animals. If need be you could resort to machinery using random numbers to keep changing the view. Growing things, whether by gardening or raising children, will help to anchor people in reality. In addition there ought to be 'something hidden beyond the horizon'. That last requirement will not be satisfied in a hollow sphere or cylinder, where the whole territory of the space settlement will be visible at the same time; on the other hand, the landscape will be satisfactorily large and eventful. The torus, curving upwards to provide a built-in horizon, will offer a compromise between the wish for wide vistas and the desire to keep something hidden. Among Maruyama's hopes, visits to other space settlements will help to avoid a sense of isolation.

The large, open structure will be suitable for many different kinds of communities which are more or less closely knit, but people who are highly individualistic, like Texan ranchers on Earth, may prefer the comparative privacy of the units of the necklace configuration. Individuals will continue to differ in their wish for interaction: some being very happy in New York City and unhappy on a farm, others vice versa. Therefore space settlements of different sizes will be needed.

To the American scientists and engineers, Maruyama was at pains to emphasise that their ideas about space-settlement design might represent a small sample of the social philosophies known to mankind. The correspondence he saw between the shape, size and social organisation of space settlements reflected more general themes from Maruyama's analysis of human societies. A space community devised by Navajo Indians, for example, might be quite different from anything that would occur to Western experts. The chief distinction that Maruyama drew was between 'homogenistic' and 'heterogenistic' cultures: meaning on the one hand the search for a 'best' way of life supposedly applicable to everyone and, on the other, cultures that valued diversity for its own sake.

Maruyama lumped European, American, Chinese, Hindu and Islamic cultures together under the 'homogenistic' heading. Despite the apparent contrasts between East and West, there is a shared logic of hier-

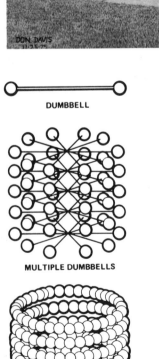

DUMBBELL

MULTIPLE DUMBBELLS

MULTIPLE BEADED TORUS

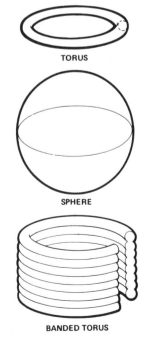

TORUS

SPHERE

BANDED TORUS

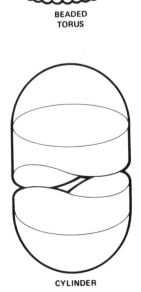

BEADED
TORUS

CYLINDER

archy and order and a belief in 'one truth'. And in the 'homogenistic' space settlement of Maruyama's imagining, there will be a clear zoning of activities, with the areas for living, working and taking recreation being quite separate. All the houses will look alike, but residents will be divided by age, occupation and so on, into homogeneous groups. There will be competition for places in the hierarchy, and majority rule. Differences will seem inconvenient and people who are abnormal will be corrected by being made more normal.

'Heterogenistic' cultures are to be found in Japan, especially in the oldest Jomon tradition, as well as in many Eskimo and American Indian tribes (the Navajo, for example) and in African tribes living south of the Sahara, typified by the Mandenka of West Africa. In the 'heterogenistic' space settlement every house will be different, rejecting the belief in a 'best way'. But, in the spirit of Japanese flower arrangement, the inhabitants will strive to achieve harmony out of diversity, avoiding competition and antagonism while mixing together people of all ages, occupations and family status. Recognising that majority decisions can inconvenience individuals, they will eliminate or compensate any resulting hardship.

Societies in different parts of the world know many subtle variations on these broad philosophies. Maruyama classified the variations, too, and saw them contributing to the admirable diversity of our species. But while he hoped that many different kinds of communities would be created in space, Maruyama made no secret of his own preference for the intrinsic diversity of the 'heterogenistic' culture. He liked to illustrate it from the philosophy of the Mandenka tribesmen: 'If you force individuals to be similar,' they said, 'the only way left for them to be different is to get on top of one another; this creates conflicts.' In Maruyama's opinion, their cultivation of diversity anticipated the findings of modern biology about the importance of generating new possibilities, and about the value of symbiotic relationships between individuals using different resources to different ends.

One may doubt whether, even in space, the world's principal cultures will change as radically as Maruyama thinks they should, and yet welcome his comments as a useful antidote to utopia-makers who pay too little regard to human diversity. And he is right in thinking that evolutionary biology illuminates the logic of diversity, and that it should teach us some unconventional wisdom about the differences between people.

Closely coupled with Western civilisation's pursuit of 'truth', or quest for the best, was the notion of an ideal type of person to which mere mortals approximated to different degrees, thereby arranging themselves naturally in social and racial hierarchies. Even in the twentieth century it continued to infect biology and anthropology, in 'mutationist' genetics and in the theories of the Social Darwinists, which had very little to do with Darwin. Modern evolutionary theory and especially discoveries coming out of molecular genetics in the 1960s demolished that notion, for anyone with the wit to see it.

The central point of Darwin's theory of natural selection was that the differences between individuals were the source of evolution, and represented a species' potential for adaptation and survival. Corrupt versions of Darwin's theory regarded differences as departures from a norm and genetic mutations as dreadful, except in rare miraculous instances where they produced a great leap forward. But such notions crumbled when molecular biology and protein chemistry enabled people to read the differences in the genes. Harry Harris in London and Richard Lewontin and Jack Hubby in Chicago found, in human beings and in fruit flies, that viable mutations – alternative versions of genes persisting in a population – were far commoner than most people had suspected. There were so many options, in fact, and so many ways of shuffling them, that no individual or group could possibly claim a near-perfect set; any idea of genetic perfection became meaningless. And this amazing amount of genetic diversity was not a peculiarity of a newly evolved species: similar genetic analysis of animals that had scarcely changed outwardly in many millions of years showed that they, too, retained a comparable degree of molecular variability – as if they were still testing possible changes in every generation, and deciding against them.

The moral is sharp and simple: living nature values diversity for its own sake. The differences between people, which reformers try to iron out, and reactionaries to exploit, are hallmarks of a sound species. Socially and biologically we need all those diverse in-

A settlement *built as a large spinning torus or doughnut* (*NASA painting*, above left) *offers a compromise between the wish for spaciousness and the wish to keep something hidden beyond the horizon – even if the hori-* *zon curves upwards. Among other imaginable shapes for space settlements* (left) *the simple sphere or cylinder might suit united societies, while the composite structures would favour privacy and independence.*

dividuals, however irritating it might seem to narrow-minded pursuers of the one best way. And the diversity embraces also the different temperaments and prejudices of people, and so allows for conflicting opinions about how life should be – logically including even a rejection of the logic of diversity!

Despite all the differences and conflicts of interest, the complicated societies of mankind function tolerably well on the whole. That would be improbable if human behaviour was guided only by the whims of individuals. The search for generalisations about human nature, which might help to describe and then to improve human societies, is not an empty hope. While archaeologists, anthropologists and social psychologists were pursuing it in their own ways, the biologists were offering a basis in evolutionary theory for a unified science of human behaviour. One of the core ideas came from William Hamilton of London University in the 1960s, but the coherent thesis about the evolution of animal and human social behaviour appeared in 1975, in *Sociobiology*, by Edward Wilson of Harvard. It aroused a great deal of controversy.

The sociobiologists looked throughout the animal kingdom for patterns of behaviour which they could interpret as having evolved in much the same way as limbs and teeth evolved, and therefore as being part of the animal's genetic make-up. The initial comparisons with other species were not unflattering to human beings. So far from being naked apes with a killer instinct they appeared, in sociobiology, to be exceptional among the higher animals in their ability to live cooperatively in large communities. People seemed less likely to murder their own kind, even taking war into account, than many other species of animals. Nevertheless, human beings were certainly not saintly in their biological predispositions and sociobiologists could claim that anyone wanting to create a better society would be well advised to take account of the constraints of human nature. Another utopian prejudice became hard to defend – the idea of the noble savage.

We invited Melvin Konner, a young Harvard anthropologist, to speak for sociobiology. Now Konner knew the San (Bushmen) of Southern Africa and dismissed any idea that those primitive hunter-gatherers were inherently happier and more peaceable than ourselves – as other anthropologists had often wanted to argue. In fact the San had a homicide rate comparable with the homicide rate in an American city and they engaged in frequent conflicts, both verbal and physical, in spite of their relatively strong sense of community and social harmony. Konner's wife, Marjorie Shostak, studied the lives of San women and found maternal conflicts, between mothers and children and between mothers and grandmothers, reminiscent of what one might hear while interviewing a woman in London or New York. Thus the common threads of human experience appeared to be stronger than was previously supposed. Sociobiological theory offered evolutionary reasons for expecting family conflicts to occur in all societies.

Some people, Marxists in particular, had spoken out vehemently against the emergence of sociobiology and linked it with reactionary ideologies. Konner thought there was no necessary connection of that kind. He liked to remind the critics that Karl Marx was such an admirer of Charles Darwin that he offered to dedicate a volume of *Capital* to the evolutionary theorist. The new evolutionary perspectives on the constraints in human social behaviour did seem to Konner to eliminate some utopian ideas, including some utopian socialist ideas, and to help explain why competitiveness and selfishness re-emerged in various socialist systems which set out to abolish them. But better understanding of the reasons could help to curb those tendencies in the future.

Comparisons of human social behaviour and social organisation in different cultures made it clear, to Konner's mind, that competitiveness and selfishness were enduring human qualities, while the broader context of animal evolution made such behaviour comprehensible. So there was no reason to believe that there could ever be a social organisation in which competitiveness and selfishness were completely eliminated. On the other hand, there were different degrees of competitiveness and selfishness in different types of human society, so that reducing them was perfectly conceivable. Similarly humans had a deep-seated tendency towards conflict, but there was ample experimental evidence suggesting that it, too, could be reduced. For example the biological approach to human behaviour provided the clear indication, consistent with observations in a wide variety of non-human species, that women were less aggressive than men. To Konner that suggested women should have more power.

An external view *of the torus-shaped space settlement visualised on page 122 appears in another NASA painting. Note the lumpy shield around the outside, to protect the inhabitants against cosmic rays. Calculations of the necessary mass of shielding suggest that an elegant shape like this is a more remote possibility than a simpler sphere, cylinder, or banded torus. Architecture will not easily amend human behaviour.*

Edward Wilson, the chief prophet of sociobiology, fielded the hardest questions. I quoted him earlier (see Chapter 4) on the ethic of botanical and zoological diversity, and on the future management of plant and animal evolution by human beings. So it was natural also to discuss with him, in this new light, the old issue: should people take charge of human evolution? Might they perhaps try to improve not individual intelligence, the usual obsession of the 'eugenicists', but social behaviour? Wilson commented:

'It is inevitable that we are going to continue to change genetically through time. If we don't try to monitor and direct that change then we are going to drift. This may be a moral judgment we wish to make in the future. But if we decide to drift then that would be making a decision about our own future – simply not to make a decision. If we decide not to drift, we are going to have to direct our own evolution to some extent . . . We are going to encounter deep pitfalls.'

Among the conceivable measures, 'cloning' would be an extreme but vivid case. It would mean generating a human being from a cell taken from a living person, bypassing the usual process of sex, so that the new individual produced by cloning would be genetically identical to the one from which the mother cell was taken. Thus it would be possible to create a large population of identical individuals, from someone thought to be genetically 'perfect'. But Wilson pointed out that the race of genetically 'perfect' people that resulted would lack all biological diversity and would be greatly imperilled in a viral epidemic, for example.

By one of the basic theorems of sociobiology, the more closely related an individual is to others around him in the population, the more likely he is to act altruistically and cooperatively towards them. So the cloned society would be strong on altruism and co-operation. But Wilson feared the other edge of that sword: 'the capacity for the group to run amuck in one direction or another, to favour one particular ideology or one extreme decision, and then to run it right to the edge of the precipice.'

To throw away genetic diversity in the human species, Wilson thought, would be to lose part of our humanity in a deep sense. We shall not understand, for several more decades at least, what the significance of that diversity really is. The human capacity for co-operation in large communities originated only during the last several million years. The remaining predisposition to distrust strangers and to compete for· status among members of the same society is, for Wilson, a holdover from our ancient mammalian heritage.

Human evolution continues, though in what direction there is no way of telling.

The further exploration of human nature may well reveal links between valuable and undesirable traits: creativeness linked with a desire to be domineering, perhaps; or the capacity for fanaticism and chauvinism. If so, attempts to sharpen the capacities for the good qualities will have the effect of augmenting the bad traits as well, and we shall come to realise that human nature has in-built inconsistencies and tensions. This was Wilson's parting shot:

'Traits such as hypocrisy and deceit and even self-deceit may be a fundamental aspect of human nature that can't be eradicated altogether. Rather than an ideal condition, a utopia of the future, we may find ourselves continuing with thousands of years of politics.'

Another biological possibility that runs counter to the logic of diversity and raises profound issues is the idea of extending the span of human life significantly. The desire for immortality, as old as mankind, received a modern fillip when people began contemplating travel to the stars and realised that the human lifespan would be short compared with the durations of the journeys. Moreover, animal experiments suggested possible routes to longevity. But Sol Spiegelman, molecular biologist and cancer researcher at Columbia University, was one of the vehement objectors. The aim should be, he thought, good health and freedom from pain during a normal lifespan.

Spiegelman drew a moral of a kind from experience of cultivating cells in 'tissue culture'. Normal human cells or animal cells of any kind had a limited lifespan. In tissue culture they would go through a certain number of divisions and would then die – the number of divisions being related to the age of the individual from which the cells were taken. But when experimenters transformed them to cancer cells, then they became immortal. Indeed the only immortal cells of animals were cancer cells. Spiegelman said:

'It's a very perverse and selfish idea to want immortality because if it were granted it would essentially mean the end of the species. The only way a species can continue to exist and evolve is to replace its individual members. So from a broad biological point of view it's wrong and therefore should not be a goal of biomedical research.'

There may, though, be ways of achieving mental immortality that do not preserve the individual's body or his genetic peculiarities. The idea of 'extending human consciousness' by meditation or the use of psychedelic drugs has been a strong undercurrent of Eastern life for millennia and of Western life for a couple of decades. In contrast with those approaches to extending consciousness the Columbia physicist, Gerald Feinberg, proposed a practical route to extraordinary cerebral possibilities. He started from firm new knowledge about the organisation of the human brain and its division down the middle into two distinct, semi-independent halves, connected by cables of nerves.

Research with patients who had the cables cut surgically as a treatment for epilepsy revealed different talents in these two halves of the brain. (I described the classic investigations by Roger Sperry and his colleagues in *The Mind of Man*, 1970.) The left side of the brain is stronger in coping with verbal and analytical tasks, while the right brain is better at seeing the 'whole picture' and dealing with relationships. In ordinary people the functioning of the two halves is thoroughly mixed. For Feinberg's first idea, one should be able temporarily to separate or suppress one of the two halves. The individual should be able to function, for a while, first according to one mode and then to the other – so experiencing the different abilities of the two halves of his brain. Beyond that lies the prospect of artificial telepathy.

Seeing that relatively small cables of nerves carry all the necessary communication between the massive brainpower of the two halves led Feinberg to the further idea of extending the linkages outside the skull. In his scheme, the brains of a number of different people will be linked by cable or radio. They will then be able to share the experiences and memories of each other's brains in a much more thoroughgoing way than speech and other ordinary means of social communication allow. For Feinberg, that novel form of consciousness will be an evolutionary step forward equivalent to going from single-celled organisms to multi-celled organisms.

Such a community of brains may have a much higher ability for thought than the sum of the individual brains. It may also possess a form of immortality. If the thoughts and experiences of a given brain are to be shared by the other brains in the community then, when it dies, its memories will live on in the survivors. They can then replace the defunct brain with another and so keep the community going indefinitely. Guessing at the subjective benefits and outward consequences of Feinberg's extraordinary scheme may be beyond the capacity of ordinary brains – some of which I have heard declaring forcibly that they have no wish to read other people's thoughts.

The interactions of human brains and computers will add new dimensions to human mental activities in the future. Reflecting on life in a space settlement the Bristol psychologist, Richard Gregory, gave his opinion that computers will help people to enlarge their imaginative space, however confined their physical environment may be. I would add that, as 'artificial intelligence' develops and computers become more manlike in their talents and responses, there will be a sharing of mental as well as physical effort. This is a major theme for any consideration of human futures; regretfully I shall not pursue it here.

Early in this book I stated the principles for a human break-out into space using the ideas of the late Desmond Bernal of the Bernal Sphere, Freeman Dyson of the Dyson Sphere, and Sol Spiegelman speaking about the extraterrestrial potential of DNA. It seems fitting to refer back to these three for their thoughts about the remaking of human beings as they move out into space. Writing in 1929 Bernal envisaged largely mechanised space people, in which the inefficient and vulnerable human body will be replaced with mechanical attachments to the indispensable brain, including an artificial 'heart-lung digestive system'. Bernal's remade human organism would have artificial senses, allowing it to receive radio, infrared rays, ultrasound and so on. It would be in intimate mental communication with other individuals (much as Gerald Feinberg has described) producing an immortal community of minds. Advances in medical engineering and surgery in the half-century after Bernal offered his description have gone a long way towards validating it, while the advent of genetic engineering raises the possibility that its techniques, too, will be applied to transforming the human organism.

But Bernal, impelled by reason to contemplate extraordinary para-human creatures, nevertheless admitted to feelings of distaste in imagining them. His humane reaction was probably the right one, and more important than all his ingenious technical thoughts. In the 1970s, the consensus among scientists was that, for the foreseeable future at least, we should reconstruct the universe to suit human beings, not reconstruct humans to suit the universe. Freeman Dyson put it like this:

'I would hope that we would leave ourselves alone. I think we can do much better by going out there and living with whatever wits Mother Nature has given us, as the Pilgrims did when they first came to America. It wasn't any great technological tricks that allowed them to survive here, it was virtues of an old-fashioned kind – common sense, good humour, ability to cooperate, ability to improvise and, of course most important from their point of view, a faith in divine providence.'

The logic of diversity and of evolution leads, though, to an uncomfortable conclusion about humans going into space. According to Sol Spiegelman, the human species will certainly divide. As it spreads in space it will split biologically as well as physically and politically. Spiegelman remarked upon a feature of the human race and primates in general – their tendency to evolve at a high rate. One reason was that they often went in small groups rather than in large herds, and that enabled genetic mutations to become quickly established through the population. Civilisation reconnected the various human groups that had spread around the Earth, and intermarriage between them made them more and more like a single breeding population. Going into space will reverse that recent trend.

The very act of making a decision to go or to stay, Spiegelman noted, will immediately select certain kinds of individuals who will opt to go. Then breeding from small groups and the quite severe natural selection in space will accentuate particular kinds of traits. Whether or not human genetic engineering accelerates the process, in time the differences will reach the point where interbreeding with the people remaining on Earth will become more and more difficult. The process will end in splitting off distinct human species – amounting to a clear division in human evolution. If the members of the new species are more intelligent and kindlier than ourselves then, as Spiegelman suggested with some irony, friendly exchanges may continue between us.

A last word on human nature as it stands at present. If anyone is unduly optimistic about the break-out into space bringing prompt maturity to our species, they should consider what happened to a message to the universe. It was prepared by American astronomers, to be carried in one of the *Voyager* spacecraft for possible aliens to intercept far beyond the Solar System. At the last moment officials of the National Aeronautics and Space Administration removed photographs of a nude man and woman, intended to give the aliens a hint about our biology. They said that the photographs were pornographic. They added instead the names of the congressional astronautics committee.

On the other hand, the urge to send a message at all, in *Voyager* as in an earlier *Pioneer*, has a special meaning, I think. The package, with its greeting from President Carter and its recording of Bach (to 'show off'), is a statement like Stonehenge. Kilroy is still here, defying mortality.

11

The nearest star to the Sun, Proxima Centauri, is more than four light-years away – that is to say, it takes its light travelling at 670 million miles an hour more than four years to cover the distance. Many of the bright stars, like those in the constellation of Orion for example, are hundreds of light-years away. But as twentieth-century astronomy defined more precisely the scope and scale of the observable universe, the human mind was left boggling at the immense distances across the void starting at the top of our heads. When they went to the Moon, men covered barely more than a hundred-millionth part of the distance to Proxima Centauri, and objects were to be seen lying a billion times farther than that star.

The Sun was interpreted as a middle-sized, middle-aged star in one of the spiral arms of the Milky Way Galaxy. The Galaxy was made out as a flattened, spinning disc of stars which we saw edge-on as the splendid river of light across the night sky: but all of the stars around us were part of the same Galaxy, which embodied more than 100,000 million stars of different sizes and ages. Old stars were dying and new stars were

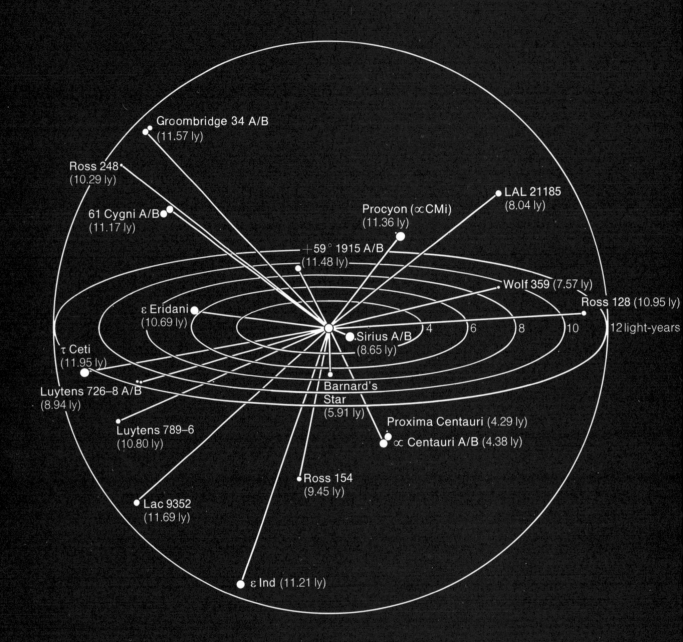

Groombridge 34 A/B
(11.57 ly)

Ross 248
(10.29 ly)

61 Cygni A/B
(11.17 ly)

LAL 21185
(8.04 ly)

Procyon (∝CMi)
(11.36 ly)

+59° 1915 A/B
(11.48 ly)

Wolf 359 (7.57 ly)

ε Eridani
(10.69 ly)

Ross 128 (10.95 ly)

Sirius A/B
(8.65 ly)

4   6   8   10   12 light-years

τ Ceti
(11.95 ly)

Luytens 726–8 A/B
(8.94 ly)

Barnard's
Star
(5.91 ly)

Luytens 789–6
(10.80 ly)

Proxima Centauri (4.29 ly)

∝ Centauri A/B (4.38 ly)

Ross 154
(9.45 ly)

Lac 9352
(11.69 ly)

ε Ind (11.21 ly)

The nearest stars *are probably the only ones that could figure in any serious programme of interstellar exploration and travel in the next century or two. All of the visible stars out to a distance of twelve light-years from* *the Sun and the Earth are shown in this diagram. Nearby stars thought to possess unseen companions – possible planets – are Proxima Centauri, Barnard's Star, Lal 21185 and 61 Cygni A.*

being born, all the time. Some of the biggest and hottest stars seemed to burn up in a few million years and destroy themselves in the cataclysmic explosions known as supernovae – leaving as residues incredibly dense agglomerations of matter: dwarf stars, neutron stars (detected as pulsars) and possibly black holes. But small stars had lifetimes of many billions of years and would go on smouldering far longer than the Sun. The diameter of our Galaxy, the scene of all this action, was about 100,000 light-years.

In the cosmic scale, even the Milky Way Galaxy was a speck: just one of ten thousand million galaxies in the observable universe, scattered like ships across ten thousand million light-years of space. The nearest small galaxies, the Magellanic Clouds, were 170,000 light-years away; the nearest large ones, M31 and M33, were two million light-years away. But the light and radio waves from the most distant objects known started their journey to our telescopes long before the Sun and the Earth came into being.

Until the twentieth century was drawing to its close, these enormous numbers and distances might prompt an interested gasp or two, but they were so great as to put them right beyond any human frame of space and time. One physicist rebuked anyone who dreamed of travelling to the stars as being like a child who wanted to feel the pictures in an art gallery. But then the astounding possibility arose that humans might actually traverse the nearer distances, and ultimately take over the Milky Way. Some people began to look at those rich starfields with quite different eyes: seeing every point of light as a new sun by which people might one day live; seeing them also as stepping stones beckoning all the way across the sky.

This chapter tells first of scientific conjectures about other intelligent beings who might be able to send signals or rockets across great interstellar distances. Then I report a serious scheme for interstellar travel by man-made robots, which could pave the way for human expeditions to other stars. The chapter ends with a reappraisal of whether those alien intelligences exist out there.

In November 1974 American radio astronomers celebrated the refurbishing of their big dish at Arecibo in Puerto Rico by transmitting a three-minute warble in the direction of a distant cluster of stars. It described briefly in code, for any being who might intercept it, our Solar System and our life on Earth. When the British radio astronomer, Sir Martin Ryle, heard about it he wrote to his colleagues saying, in effect, 'Don't do that sort of thing.' Ryle set about trying to persuade the

world's astronomers to agree not to advertise our existence to other civilisations; at the very least he wanted full international agreement on that possibly fateful step. An expanding galactic civilisation might be looking for new comfortable planets to settle on, to farm or to mine. By luring potential colonisers from far away in space we could bring about our own destruction, or suffer piracy and enslavement on a global scale.

It was a sign of how deeply into the scientific consciousness the notion of a habitable and traversable universe had cut, that such apparently far-fetched notions should come from someone of Ryle's high standing in the astronomical community. He was not just a prominent practitioner but one of the founders of radio astronomy, and the ingenious combinations of telescopes which he and his colleagues at Cambridge used for a quarter of a century, to map the radio universe, won him a Nobel prize. Unlike his sparring partner, Sir Fred Hoyle, Ryle had no reputation as a story-teller. And so far from being an enthusiast for relentless technology, he was conspicuous in the anti-nuclear lobby, advocating windmills in preference to fast breeder reactors. Yet technology might carry you among the stars. As Britain's Astronomer Royal, Ryle took a position very different from that of his unfortunate predecessor, who never lived down an incautious declaration, 'Space travel is utter bilge'.

Among Ryle's fellow astronomers, especially in the USA and the USSR, the search for extra-terrestrial beings became a practical enterprise after 1959, when Giuseppe Cocconi and Philip Morrison had formally set out the possibility that alien intelligent beings might actually be trying to get in touch with anyone who was listening. By the mid-1970s, the US National Aeronautics and Space Administration's Ames laboratory was studying ideas for enormous radio telescopes, on the Earth or in space, intended to detect coded signals from the universe. For an up-to-date assessment of the prospects, Dick Gilling and I turned to Frank Drake of Cornell University, the man who sent the signal that worried Ryle.

In 1960 Drake had been the first person ever to point a radio telescope at nearby stars, in the hope of detecting intelligent signals. I had met him later at Arecibo, where he was running the largest radio dish in the world, a thousand-foot telescope. At that time, in 1968, it was busy in the great pulsar hunt. When Jocelyn Bell and Anthony Hewish, in Ryle's group at Cambridge, had detected the suspiciously regular radio bleeps from the first pulsar, they had labelled them LGM for 'little green men', until they satisfied themselves that the

Alien intelligences *in the universe may be detectable by radio. Already active in the search is the radio telescope at Arecibo in Puerto Rico* (above), *which has a 1000-foot dish slung in a natural crater. It could, in principle, communicate with an instrument similar to itself, on another planet, out to a range of thousands of light-years. The searchers for extraterrestrial intelligence would like in the long run to construct a similar but even larger radio telescope in space* (model, right). *It might be a dish as much as two miles in diameter, shielded from terrestrial radio noise by a metallic screen.*

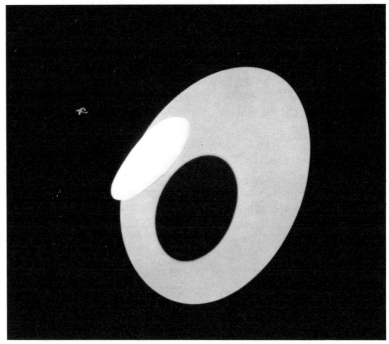

signals had natural origins in a highly peculiar star. For Drake and his young men at Arecibo, looking for pulsars was not at all dissimilar from the search for intelligent signals against the background of natural radio noise from the universe.

Drake contended that the choice of means for interstellar communication would be determined, not by the laws of physics or even the arrangement of the universe, but by cost – primarily the cost in energy. Rocketry was, he thought, a preposterously expensive means of searching among the stars, while the use of electromagnetic waves was far more economical. One could deduce the precise radio frequencies which would be best suited for interstellar communication and discourse. They were frequencies for which radio astronomers on Earth were already well equipped, and which could therefore be used to search for signals coming from distant worlds. As Drake put it:

'We expect the interstellar messengers not to be flaming rockets but to be invisible radio waves, waves which may be sweeping through this room right now. And this is true for us and for every civilisation. For that reason we believe it is most promising to search the heavens with large radio telescopes for the . . . transmissions of other civilisations.'

In his first attempt, Drake used a radio telescope at the US National Radio Astronomy Observatory (NRAO) in West Virginia. He searched in vain for intelligent signals from the vicinity of two of the nearest stars resembling the Sun, Tau Ceti and Epsilon Eridani. By the mid-1970s the pace had greatly quickened. NRAO instruments had scanned about a thousand stars for signals. Two groups in the Soviet Union undertook a long series of searches, listening simultaneously with radio receivers at five locations. Radio astronomers at Algonquin Park, Canada, at the Ohio State University, and at Drake's own Arecibo observatory were all engaged in intermittent work on possible extraterrestrial intelligences.

The search at Arecibo turned to possible signals from other galaxies. Although they lay much farther away than the stars of our own Galaxy, you could imagine that the brightest civilisations might not necessarily be the nearest. Meanwhile the tools for the search were becoming more effective, with several thousand frequencies being monitored simultaneously and bigger telescopes and more sensitive receivers coming into use. The Arecibo telescope, for example, duplicated in one tenth of a second the work of two months in 1960.

The next step, foreseen by Drake, will be to build a million-channel radio receiver and then a billion-channel system. In the early 1980s a 100-foot radio telescope may be launched into space, to join in the search for extraterrestrial intelligent signals. Beyond that, there may be larger American efforts. Code-named Cyclops, they may include a vast array of 300-foot radio telescopes on the Earth, giving altogether hundreds of times the collecting area of any previous radio telescope. Alternatively a single spherical reflector, more than two miles in diameter, may be launched into orbit. Protected from the radiation of the Earth by a huge metallic shield, it would focus radio energy on to free-flying spacecraft positioned by laser beams. The costs of such enterprises will be measured in thousands of millions of dollars but that prompted Drake's comment: 'The cost of learning from other civilisations is a fraction of the cost of a very small war.'

Some critics, though, reasoned quite differently from Drake, saying that there was no need to build such large and expensive radio telescopes in order to detect other intelligences in the Galaxy. If they existed at all, their representatives would very likely be in our vicinity. The argument was as follows. It looked as if human beings would be capable of travelling to other stars by about the twenty-second century. Thereafter they would, in principle, be able to spread right across the Galaxy in a time very short compared with the ten billion years of the Galaxy's existence. Presumably any other technologically advanced civilisation would be capable of the same action and one or more of them would probably have swept through this part of the Galaxy.

They would know the Earth as an inhabited planet and would probably be keeping an eye on us. In 1977 Tom Kuiper and Mark Morris of the California Institute of Technology suggested various reasons why an alien outpost in the Solar System might not make itself known to us. The visitors could be waiting for us to achieve a higher level of knowledge for ourselves before exposing us to 'culture shock'; meanwhile they might be looking upon the Earth as a kind of nature reserve, not to be tampered with. Kuiper and Morris concluded that the best chance for spotting extraterrestrial intelligence at work might be to intercept communications signals beamed from the direction of the parent star to the alien outpost. The existing radio telescopes on Earth would, they thought, be adequate for that purpose.

But shall we ourselves really be able to travel to the stars? Robert Goddard, the American rocket pioneer, always thought so, although he left his ideas unpublished for fear of ridicule. Desmond Bernal, Freeman Dyson, Gerard O'Neill and other likeminded physicists did not hesitate to embrace the possibility, as being compatible with the laws of nature. Science-fiction writers took over the idea but not the constraints. They swept aside the very physics on which feasibility rested, making light of interstellar journeys which the physicists, by contrast, foresaw as taxing human skill and will to limits of heroism and endurance. In reaction against 'cereal box' notions of space travel, some physicists did sums proving to their own satisfaction that interstellar travel was physically possible perhaps, but economically quite impossible. By the 1970s that pessimism was being challenged. The human bridgehead in the Solar System, once it begins to multiply the material assets and energy supplies of our species, will open the way for interstellar travel to begin.

Here my chief mentor has been the engineer Alan Bond of the British Interplanetary Society. Bond always had his eye on space: he began his career working for Rolls-Royce on the Blue Streak rocket, which was intended to be the booster for a European space launcher until the politicians stopped it. When I met him he was at the Culham Laboratory, as a member of the team planning the Joint European Torus for thermonuclear fusion research. Bond enjoyed a high reputation among the younger generation of British engineers and his polymathic interests extended to astronomy and the origin of intelligent life. In his studies of interstellar propulsion by nuclear-fusion rocket his closest associate was Anthony Martin, also at Culham, but they were both careful to point out that this was spare-time work, and nothing to do with their governmental employers. It amounted, nevertheless, to the first detailed description of how the escape from the Sun could begin.

In the 1930s the British Interplanetary Society produced a study of how to travel to the Moon – at the time nearly everyone thought it was crazy. In 1977 a group of a dozen scientists and engineers led by Bond completed a five-year study of corresponding craziness, of a possible ship that could leave the Sun behind and visit the planets of one of the nearest stars. Project Daedalus they called it, but they knew quite well that they could not expect to see the necessary monstrous vehicle built in their lifetimes. And the craft they envisaged would spend the best part of some other human lifetime on its immense journey. It would be unmanned and could not return to Earth, nor even slow down near its target star. Nevertheless, it could be the forerunner of other ships carrying people far beyond the Solar System. Earlier I mentioned Bond's view of the nuclear-fusion engine as the propulsion system of choice for high-speed spaceflight; also the advantages of the rare light form of

helium as a fusion fuel producing relatively few neutrons – even though people will have to fetch it from Jupiter. The interplanetarists brought these possibilities together in the design of their starship.

Imagine *Daedalus* being assembled, perhaps a hundred years from now, in orbit around Callisto, the moon of Jupiter. It will be a great structure, comparable in size and mass with an ocean liner – 600 feet long and 54,000 tons – but looking more like a Byzantine church. Most of the bulk will consist of pellets of thermonuclear fuel. The payload will be 400 tons and the energy expended in dispatching it will be equal to about four years' production of energy on Earth by the entire human species in the 1970s. By the time people construct the first starships the terrestrial energy problems will, by implication, be fully solved.

Daedalus will be sent on its way by thermonuclear explosions produced by compressing and heating pellets of heavy hydrogen (deuterium) and light helium (helium-3). High-energy electron beams will converge on each pellet in turn, zapping it into thermonuclear reaction, in accordance with a power-producing technique already under test at American and Russian laboratories. The pellets will go off at a rate of 250 a second, roughly the rate of fuel explosions in a car engine. As in lesser rockets, there will be an advantage in dividing the vehicle into stages, so that the craft sheds unnecessary structural mass as its journey proceeds. The first stage of *Daedalus* will run for two years and then fall away, leaving the second stage to operate for a further twenty-two months. By that time the ship will be travelling at 24,000 miles a second, or about thirteen per cent of the speed of light. Thereafter it will coast to its destination.

Such enormous speeds create problems of damage. Even in the vacuum of interstellar space traces of dust will, according to Bond, erode away some sixty tons of the vehicle's structure during the flight. Near to the Sun and the target star there will be much more abundant and dangerous flotsam. While *Daedalus* will leave the vicinity of the Sun at a relatively low speed, special protection will be needed as it approaches the target star, in the form of a shield of debris that will be flown ahead of the main vehicle, to clear its path.

'To say that the craft would be unmanned is a relative term,' Bond told me. The payload will include a very big computer and two robots called Wardens. Assuming plausible advances in electronics and artificial intelligence, the Wardens will be extremely clever machines, capable of a great deal of reasoning and independent decision-making – very necessary when messages to and from the Earth will take years to pass. During the

Barnard's Star *(upper photographs) is the target proposed by the British Interplanetary Society for an unmanned starship* Daedalus *(model, lower illustration). This relatively faint star is one of the closest to the Earth – about six light-years away. The successive pictures show Barnard's Star moving in relation to more distant stars. During the past forty years, its track has perceptibly 'wobbled' under the influence of unseen planets.* Daedalus *would be a 54,000-ton craft propelled by a rapid succession of small nuclear explosions. The large spheres would contain the fusion-fuel pellets of the first stage; a smaller second stage is shown nestling among them. Even travelling at 90 million miles an hour,* Daedalus *would take half a century to reach Barnard's Star.*

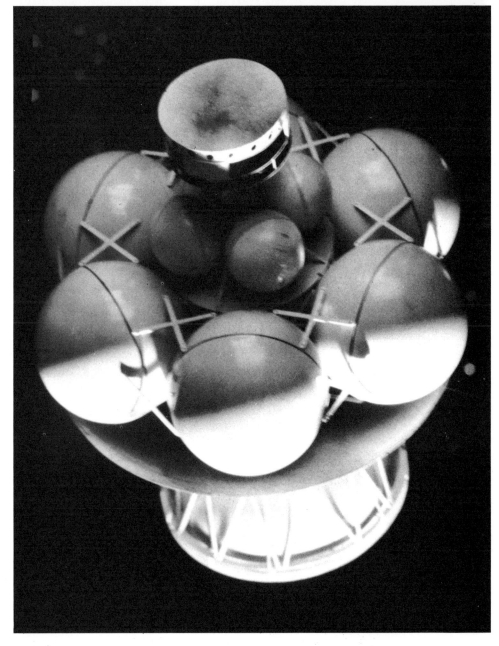

long coasting phase, several probes will fan out thousands of miles from the ship, clear of its pollution, to relay information about the interstellar dust, gas and radiation. All the time, the ship will be reporting back to the Solar System. The propulsion chamber, its main work done, will become a radio dish for that purpose.

The target star chosen for consideration by the interplanetarists was Barnard's Star, lying about six light-years away from the Solar System – a small red dwarf, which astronomers suspected of having planets in orbit around it. The flight of *Daedalus* to Barnard's Star will take fifty years and the instruments on board will be observing the star closely during the last years of the approach. But the busiest time, naturally, will be during the twenty hours or so when *Daedalus* is passing among the planets of Barnard's Star, before hurtling into the void beyond. Big telescopes and swarms of independent probes released by the ship will harvest an enormous amount of unpredictable information – so much that it will take about a year to relay it all back to Earth.

Not that it will necessarily be quite like that. Bond was the first to insist that *Daedalus* was only an imperfect paper study. Before the first interstellar vehicle is built, in the twenty-first century, technological ideas will have evolved beyond that spaceship of the mind. There have been, for instance, various conjectures about using the very tenuous gas and dust in the spaces between the stars as a source of fuel or ejectable mass for the propulsion of starships. Nevertheless, the *Daedalus* study convinced Bond that in due course man-made vehicles will indeed be travelling to Barnard's Star and other nearby stars. And he declared:

'Who knows, a short while after that, maybe men will follow if the way is known. However, it will be a one-way trip. No foreseeable propulsion system would enable anyone to make a two-way journey within the lifetime of one or even two generations of human beings. Eventually, of course, human beings may spread across the Galaxy . . . something like mould on a wall. The colonies at the outermost fringes themselves will be establishing colonies, and so the process could go on.'

Unlike the Wardens in the *Daedalus* probe, the occupants of a manned starship will wish to stop when they reach the vicinity of another star. Unless 'scooping' of interstellar gas becomes feasible, this requirement for stopping will mean an extremely heavy penalty in the launching mass of the starship. Roughly speaking it remultiplies the ratio of the launching mass to the payload, which in the case of *Daedalus* will be about a hundred. So a 'stoppable' *Daedalus* would have to be a hundred times more massive at its launching. In prin-

ciple, the overall mass ratio might be less – perhaps a hundred times less.

Travel to the stars will be much easier, safer and more congenial for the people concerned if the starship can be huge – a roving version of one of Gerard O'Neill's settlements, accommodating thousands of people in spacious living conditions that they can tolerate for a lifetime. Indeed, for the ecological reasons pointed out by Lynn Margulis (see Chapter 4) this spaciousness may be necessary for survival. But such a settlement would, with suitable shielding, have a mass of the order of a million tons. Using the fusion of deuterium and helium-3, as in *Daedalus*, and with the requirement of stopping, you would finish up with a starship whose mass, at launch, was between 100 and 10,000 million tons, or ten thousand to a million tons for every person on board.

O'Neill himself speculated in *The High Frontier* about the use of anti-matter as a fuel. This is the most efficient fuel imaginable because its entire mass converts itself into energy by annihilating itself in reaction with an equal mass of matter. Anti-matter is difficult to make and extremely dangerous to handle, but O'Neill imagined frozen anti-hydrogen (each atom consisting of an anti-proton and an anti-electron) being stored in electric fields within well-shielded containers. I estimate that, using anti-matter for propulsion, a million-ton starship payload that is to travel at ten per cent of the velocity of light and stop at its destination will need a mass of fuel (half matter, half anti-matter) of a few hundred thousand tons.

Such numbers appeared utterly preposterous in the claustrophobic world of the 1970s, apparently overcrowded, short of energy and threatened with general impoverishment. But, according to the ideas rehearsed earlier, the resources of the Solar System will support far larger populations than ours in far greater prosperity than most people at present have on Earth. I have described Theodore Taylor's Santa Claus Machine and Freeman Dyson's self-reproducing version. Given the ceaseless energy of sunshine and the potential of nuclear fusion in space, there will be no early limit to human activity. And the correct way of assessing the value of a manned starship may be to ask: 'What would you pay for the Sun?' A successful expedition to colonise the environs of another star will be in effect buying a new sun for the human species. Will a billion billion dollars be too high a price?

But what about encounters with those alien intelligences? The famous physicist Enrico Fermi once asked a famous question: 'Where is everyone?' By that he meant: if there were other intelligent beings in the universe, why didn't we see them in our vicinity? Despite the claims of UFO spotters and alleged visits to the Earth by spacemen in ancient times, there was no sound evidence for their existence. When Bond and his colleagues in Project Daedalus had convinced themselves that interstellar flight was indeed practicable for an advanced civilisation, they considered another possibility just as amazing as the idea of alien intelligences: that we ourselves were the only intelligent life in the Milky Way Galaxy.

Here I must admit to being predisposed to that opinion. Part of the pay-off from twentieth-century science was a thoroughgoing revision of the story of creation, from the origin of the universe, through the origin of the chemical elements in exploding stars, to the formation of the Sun and the Earth and the origin of life on Earth. There was an ever-more plausible account of how life sprang up from combinations of molecules that arose spontaneously in a chemical soup on the young planet. The seemingly 'inevitable' origin of life and its continuous evolution from the simplest bacteria, through worms, fishes and shrews to human beings, encouraged many physicists and astronomers to think that intelligent life must be common on other Earth-like planets in the universe. But deepening knowledge of the history of life on Earth suggested a contrary view.

An extraordinary succession of fortunate circumstances were involved in the production of human beings. I shall pass over particular evolutionary inventions like the origin of blood or backbones or hands, which might imply a prejudice in favour of intelligent beings looking like ourselves, and mention only some strategic issues. For fully half the long history of life only single-celled organisms, microbes, existed on this planet. Following the ideas of James Lovelock and Lynn Margulis one might say they were busy establishing the complex cycles of Gaia. If the genetic means had not arisen for constructing multi-celled plants and animals, the microbes would still rule the Earth and intelligence would not exist. Later, after the appearance of complex marine animals and the gradual transfer of life to the land, reptiles came to dominate the Earth, culminating in the dinosaurs which prospered for more than a hundred million years. At that time our mammalian ancestors were small and only the mysterious demise of all the dinosaurs about 70 million years ago opened the way to the evolution of larger and more intelligent mammals like the apes. Cosmically, the dinosaurs represented the probability of long-lasting success on some basis other than intelligence.

Apes would not have evolved into upright-walking human beings if the tropical forests where they lived had not been thinned out and reduced to grassland. That was a result of particular movements of the wandering continents which cooled the whole world and reduced the rainfall in many places. Even then we might have remained pint-sized people with pint-sized brains, running around in small groups on the African savannah, had it not been for the rapid succession of ice ages. Beginning some three million years ago, the drastic changes of climate associated with the ice ages accelerated human evolution.

Reflecting upon this story, I became doubtful about the supposition that intelligent communicative life was commonplace in the universe. There might indeed be plenty of planets like the Earth, many of them with some sort of life on them. But the chances of a comparable sequence of events occurring pat, and throwing up intelligent life, seemed remote. For these reasons I was not really surprised to find the British interplanetarists reaching the same conclusion, by a more mathematical approach to Fermi's paradox. As Alan Bond reported:

'One can do a calculation. It's something akin to rolling the dice millions of times once every few thousand years and asking: what is the probability that a certain sequence of numbers will come up? Evolution progresses by small mutations and only . . . the ones that are fitted for development and survival are selected. The human being represents a certain route in which a certain number of selections from a random casting has occurred. If we put all the numbers into this calculation we find out that the probability of the human route arising is a very, very small probability indeed: and it's only because there are so many planets available that even once has it occurred, namely here. It may be, of course, that a few galaxies away the same process has already occurred again . . . My guess is that we're probably the only intelligent species in this Galaxy at present, and maybe ever.'

But that was a minority opinion in the 1970s, when many astronomers still thought there might be a million civilisations flourishing in the Galaxy, in which case a confrontation will eventually occur. There will be good reason for wariness on both sides. Given the dark side of human history, we know *we* are not to be trusted in encounters with aliens and perhaps, for their part, lusty 'men of action' are more likely than saintly philosophers to have populated the Galaxy on a large scale. So the case for looking out for extraterrestrial intelligences will be stronger than ever but the motive may become inverted – not to find them, but to reassure ourselves they are not there.

12

**TAKING OVER
THE MILKY WAY**

A vehement spokesman for the most ambitious of all ideas was Harlan Smith of the University of Texas at Austin. I had first heard of him as a young astronomer who, in the mid-1960s, made a discovery far beyond the confines of our Galaxy. He saw that the new-found quasars could fluctuate rapidly in brightness – decisive evidence that the quasars were not only extraordinarily strong sources of energy but extraordinarily small as well. Besides his interest in those remote phenomena among the distant galaxies Smith also observed the atmospheres of the planets and he taught his students to

look at the nearer universe as an arena for human action. When Dick Gilling and I visited him, he commanded our attention by his emphatic sense of purpose about possibilities which for others were cold conjectures. He told us:

'The universe is an incredibly rich place. It has almost more places where life could exist than there are words in the language to describe. But, the chances are, most if not all of these are barren of life until somewhere, somehow, it becomes seeded. And one of the tasks of the human race is to carry what we hope is the blessing of life to the places in the universe where it can be sustained . . . We have the opportunity to have countless descendants and for them to enjoy the blessings and mysteries of life. We also have the opportunity to turn off that challenge by insisting on regarding the Earth as the centre of the universe.'

Smith feared we might miss the chance. We had the ability to break the chains of gravity tying us to the Earth. But who knew how long this opportunity would last, or how long the human race would continue to keep the high technology and surplus wealth to be able to make the step into space? There were enough doomsday scenarios, one or more of which might work, whereby mankind would be prevented forever from taking the step – and so be left living in a constrained environment with overpopulation, pollution, and the exhaustion of energy and material resources. The troubles to be seen around us might grow to overwhelm us, either in a creeping fashion or in a disaster such as nuclear war.

Provided we do make the first step out into the Solar System we shall come, as Smith asserted, into a realm unlimited for all practical purposes in energy, material and room. There we can accommodate an almost unlimited number of human beings, if the race should choose to expand in that sense, and give a very high standard of living to everyone. We can also develop the wherewithal to make the next step, out to the stars. Most of the stars in the Milky Way Galaxy are stupendous nuclear furnaces that will supply the basic energy for unimaginably abundant life.

Once that second step has been taken, in Smith's view, it will become hard to imagine any problem or threat which could extinguish our species. To colonise space will be to take out insurance policies. With humans living in an immense variety of widely separated environments, whatever catastrophe might occur in one place, it surely will not occur in others. If the human species accepts the challenge, it has the opportunity to become perhaps immortal, for as long as the universe lasts. Smith found it hard to imagine anything other than an encounter with another civilisation that would stop the expansion of life filling the Galaxy.

It seems to me at least as futile to try to picture exactly how people will take over the Milky Way as for someone in the eighteenth century to describe intercontinental aviation. The important thing is the physics that says it can be done – that nuclear fusion if nothing else can propel starships. Given that and the mobile space settlements of Gerard O'Neill's imagination, the question of precisely what planets exist in the vicinity of

other stars is not at all critical, as long as there is something in orbit to provide the elements of life, along with constructional materials and fuels. Creating the inhabitants is the least of the problems. Given its capacity for reproduction, our species could in principle people every star in the Milky Way with a million times the present population of the Earth, in just a few thousand years. But reaching the stars would take much longer.

At each stage of the expansion, people will presumably settle down in the vicinity of the latest star, to build a new civilisation there. Even the restless ones who talk of moving on may well wish to send off unmanned probes to help decide upon a new star; then they may dispatch self-reproducing Santa Claus Machines and seeds of plants, to prepare comfortable living conditions in advance of the next migration. Freeman Dyson suggested that the seeds of giant trees, developed by genetic engineers and capable of growing on comets, might float freely and independently through space, and take root where they could. 'This might become in fact a wave of green leaves spreading through the Galaxy,' Dyson said.

The chief factor determining the rate of human expansion through the Galaxy will probably not be the speed of the ships, but the breath-catching period between the colonisation of one star and the departure for the next. Tom Kuiper and Mark Morris of Caltech assumed a 'regeneration time' of 500 years after each step of ten light-years, and a travel speed of ten per cent of the speed of light, to arrive at a figure of five million years for a technological civilisation to populate the Galaxy. The least time is likely to be about a million years: with assumptions of much more sluggish advance, the process could take 500 million years. Whatever figures you fancy for the time needed to take over the Milky Way, they will be very short compared with the age of the Galaxy and very long compared with the lifetime of a human individual.

Notice that none of the more peculiar methods of the science-fiction writers for interstellar travel have been invoked here. We have not proposed putting astronauts in deepfreeze to survive a long journey, or using black holes either as a propulsion system or as a route for diving into another part of space and time. Nor have we envisaged travel faster than light. Either to deny the possibility of amazing tricks being invented by our descendants, or to stake the future of mankind on them, would be rash. But within the compass of firm physics it is worth mentioning the almost magical consequences that would come from travel closer to the speed of light.

Mental spaceships can easily go as fast as light, even

if real ones may never quite attain that speed. Of all the physicists who ever travelled about the universe in their imagination and coolly encompassed on a piece of paper 'the caverns measureless to man', Albert Einstein was the giant among giants. He wondered what it would feel like to travel on a beam of light, to flash signals between passing spaceships, or to fall weightlessly in a runaway room through the chasm of space between the stars. From such childlike thoughts flowed Einstein's theories of relativity. His equations gave the key law of creation, concerning the interchangeability of matter and energy, while the curiously warped geometry which he introduced into the physical universe of space and time displaced Newton's theory of gravity and implied the existence of black holes. Although he died in 1955, two years before the *Sputnik*, Einstein's will remain a name to conjure with, to the farthest reaches of the Galaxy. One reason is that he saw that action could tame time.

One of the earliest implications of relativity was that high-speed motion did funny things to clocks, and travel kept you young. If a person went off on a space journey, leaving his twin brother on Earth, he would find on his return that his twin had aged more rapidly than he. While ten years might have elapsed for the stay-at-home, the traveller would think that only five years (say) had passed – provided he made his journey at a significant fraction of the speed of light. Physicists later found that radioactive sub-atomic particles travelling at high speed broke up less rapidly than when they were at rest. The simplest example came with the heavy electrons, or muons, in the cosmic rays raining on the Earth's surface. The muons were created high in the Earth's atmosphere and very quickly decayed into ordinary electrons. Were it not for the fact that, from the muons' point of view, time passed more slowly because of their high-speed motion, they would not survive long enough to reach the ground. Their appearance in detectors at the Earth's surface was continual evidence that Einstein's idea was correct.

An international team at the European Organisation for Nuclear Research (CERN) reported in 1977 that they had kept large numbers of muons circulating in a 'storage ring' in the laboratory, to make the most precise check yet on the relativistic protraction of time. Because the particles were going around in a circle, constrained by magnets, they imitated rather closely a round-trip journey in a spaceship. The outcome was that the lifetime of muons travelling at 99.94 per cent of the speed of light was almost thirty times longer than for muons at rest, in close accordance with the predictions of Einstein's theory. If a person could spend his lifetime making an equivalent journey he would appear to live 2000 years, as judged by the calendars on Earth, while to him it would seem like only 68 years.

If he were able and willing to go on accelerating ever closer to the speed of light (he could never pass it) he would gain advantages even greater than those achieved by the muons at CERN. Unimaginably large amounts of energy would in practice be necessary to accelerate starships to speeds near to that of light. Every atom that the starship encountered on its journey would become, from the starship's point of view, very energetic atomic radiation, threatening to destroy men and instruments alike. Every grain of dust would hit the craft like a bomb. Even the very cool radio energy that pervades all space, as a hangover from the Big Bang, would be transformed into a dazzling inferno. Yet, in principle, if a spaceship ever could be accelerated indefinitely at one Earth gravity – in other words, if the crew experienced just the same acceleration as people feel on the Earth's surface – it would traverse the radius of the known universe, some ten billion light-years, in less than a normal human lifespan. It will almost certainly never happen but there is philosophical comfort in knowing, as Einstein did, that if we could ride on a beam of light time would stand still. In myth, at least, we can imagine a heroic spaceman who finds the universe an entirely compact place.

The French anthropologist Claude Lévi-Strauss has described myths as machines for the suppression of time. Holding the minds and the efforts of many civilisations to the vaguely defined and lengthy programme of taking over the Milky Way will require that people weave through the scientific knowledge and the technological proposals a thread of purpose, to keep the cosmic dream alive in the stressful waking hours. Isaiah struck the right note:

'For, behold, I create new heavens and a new earth: and the former shall not be remembered, nor come into mind.

'But be ye glad and rejoice for ever in that which I create: for, behold, I create Jerusalem a rejoicing, and her people a joy.'

For a very different kind of mythmaking an American educational sociologist, Arthur Harkins, cited *Star Trek* to us, and the extraordinary attention which that television adventure series, set in a roomy starship, attracted from young people. They took to assembling for large 'trekkie' conventions. *Star Trek*, in Harkins' opinion, was a twenty-first-century analogue of the Hollywood Westerns set in the nineteenth century, and offered the same sense of expanding frontiers. In view of the extraordinary success of the movie *Star Wars*, it seems plain that older romantic themes of good and evil are easily translated into the space setting.

Life evolves and preserves a genetic message; human cultures evolve to protect and pass on the collective experience and knowledge of people. Both biologically and culturally we can sense a compelling wish to insure humanity against eventual destruction. The American writer Ray Bradbury expressed it in conversation with the Italian reporter Oriana Fallaci who tells us that, with head lowered and eyes half closed, he uttered it like a prayer:

'. . . Don't let us forget this: that the Earth can die, explode, the Sun can go out, will go out. And if the Sun dies, if the Earth dies, if our race dies, then so will everything die that we have done up to that moment. Homer will die, Michelangelo will die, Galileo, Leonardo, Shakespeare, Einstein will die, all those will die who now are not dead because we are alive, we are thinking of them, we are carrying them within us. And then every single thing, every memory, will hurtle down into the void with us. So let us save them, let us save ourselves. Let us prepare ourselves to escape, to continue life and rebuild our cities on other planets: we shall not be long of this Earth: And if we really fear the darkness, if we really fight against it, then, for the good of all, let us take our rockets, let us get well used to the great cold and heat, the no water, the no oxygen . . .'

The human sense of time will need adjustment in a practical way. The British Interplanetary Society made it a criterion for the flight of *Daedalus* to the stars that participants in the project might live to see the signals coming back. But Christopher Columbus never found the western route to Asia. The man who first sketched the possibility of putting an artificial satellite into orbit about the Earth, Isaac Newton, did not live to see the trick done. Albert Einstein did not extend his own life-span by relativistic space travel. The bolder the dream, the more probably must the dreamer leave it to be fulfilled by others.

Compare the shoeshine man and the forester: the first works for a minute or two and gets prompt payment for a gleam that will not survive the next shower; the forester, on the other hand, plants trees that take so long to grow that he cannot live to harvest them. With the industrial revolution, forestry went out of fashion. Most of us, in the countries where it happened, began to live like shoeshine men. 'Better fifty years of Europe than a cycle of Cathay,' Tennyson declared. But people will have to learn patience again, if they are to take over the Solar System and then the Milky Way. It will mean planting many trees, both literally and figuratively. There is no point whatever in rushing to be at Titan by Tuesday. In a project that will take a hundred thousand generations each mortal individual will witness only a very brief and local slice of the action, like one day's panels in a long-running newspaper strip-cartoon. Some of our grandchildren may be working among the outer planets and some of their grandchildren, perhaps, will travel in a starship. After a thousand years the sphere of life will still be very small. Progress across the sky will seem slow and the conservative routines of everyday life will predominate. Each generation will regard its circumstances as normal, just as we take farm produce and jet planes for granted. Only a minority will be busy with new marvels of spaceflight and new adventures into the unknown.

For the decades immediately ahead the only reason for not being too dilatory will be Harlan Smith's – the wish to see the space enterprise well launched before any catastrophe can prevent it. Perhaps a revival of cautious optimism about the long-term future will help to avert disaster on Earth – who knows? In the millennia to come the expansion of life into the universe can proceed at whatever pace seems congenial for the people concerned, according to the state of their technology and their attitudes to upstarts who canvass big ideas. Yet our own role remains a privileged one. Because we live at a special moment of scientific history our generation is able to launch spaceships of the mind that may serve as pathfinders for millions of years.

Index